14.95

Tuning In to RF Scanning
from Police to Satellite Bands

Bob Kay

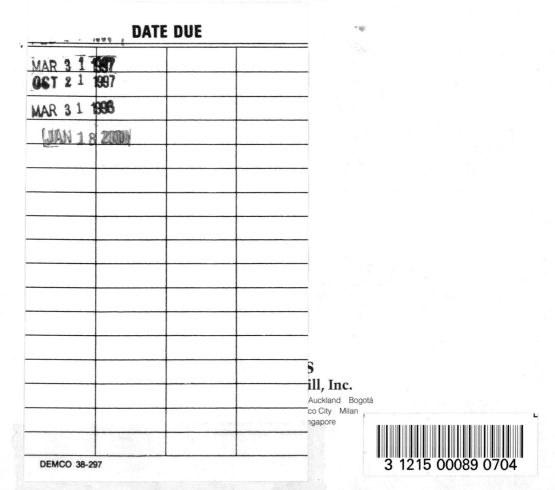

ill, Inc.
Auckland   Bogotá
co City   Milan
ngapore

1 2 3 4 5 6 7 8 9 0   DOH/DOH   9 9 8 7 6 5 4

Library of Congress Cataloging-in-Publication Data
Kay, Bob.
    Tuning in to RF scanning : from police to satellite bands / by Bob
Kay.
        p.   cm.
    Includes index.
    ISBN 0-07-033963-5 (h)   ISBN 0-07-033964-3 (p)
    1. Radio—Amateurs' manuals.   2. Radio—Monitoring receivers.
3. Radio stations.   I. Title.
TK6553.K317   1994
621.384—dc20   94-3088
                                                                          CIP

Acquisitions editor: Roland S. Phelps
Editorial team: Andrew Yoder, Managing Editor
                        John T. Arthur, Book Editor
Production team: Katherine G. Brown, Director
                        Tina M. Sourbier, Coding
                        Ollie Harmon, Coding
                        Patsy D. Harne, Desktop Operator
                        Joan Wieland, Proofreading
                        Joann Woy, Indexer
Design team: Jaclyn J. Boone, Designer
                        Brian Allison, Associate Designer
Cover design: Graphics Plus, Littlestown, Pa.
Cover photograph: Bender & Bender, Waldo, Oh.   EL1
Cover copywriter: Cathy Mentzer   0339643

Dedication
To: Robert G. Chase—a true friend.

Contents

Introduction

Scanning is a high-tech, fast-paced, action-packed hobby that will keep you on the edge of your seat. By merely pushing a few buttons on a scanner radio, you can instantly listen to your local police, fire, and ambulance departments. Push another button, and you're listening to the distress call from a sinking ship. Select another button, and you're listening to the voice of a pilot making an emergency landing. Tired of Earthly adventures? No problem. Strap yourself into your armchair and start the countdown: Five, four, three, two, one....BLAST OFF! Welcome to outer space. From the safety of your favorite chair, you can hear voice communications between the Space Shuttle and NASA.

The technology of the 1990s has truly placed the world at your fingertips. Within a matter of seconds you can tune in everything on the ground, in the air, or on the sea. The number of radio communications that you can monitor are enormous. There are billions of action-packed opportunities out there, just waiting for you to tune them in. As you read this, invisible radio communications are passing through this book and are piercing your body. You can't feel them or see them, but you can capture them.

To capture the radio signals in your area, you'll need this book and a "scanner radio." In the past, people often referred to them as "police scanners." In their infancy, scanner radios were controlled by individual crystals. Each crystal represented the numerical value of a frequency. To put it simply, you needed a crystal to listen to your local police. Since most people didn't know the exact frequency of their local police, they usually asked the salesperson for a "police crystal." Hence the name, "police scanner."

Today's synthesized scanner radios do not use individual crystals. The new scanner radios are controlled by a microcomputer chip that can be manipulated via the keypad. Best of all, you don't need a degree in electronics to use a synthesized scanner radio. If you can withdraw money from an automatic teller machine, you already possess the skills that are necessary to enjoy the hobby of scanning. To enter a frequency, all you do is press the keypad with your finger. It's simple, quick, and anyone can do it.

The key to unlocking the airways is knowing the specific frequency to monitor. Again, it's similar to using an ATM machine. Punch in the numbers and the machine instantly responds. It's the same in scanning. Punch in the frequencies and your scanner radio will instantly reward you with hours of listening enjoyment.

With a scanner radio you can easily capture and listen to thousands of radio frequencies in your neighborhood. Day or night, weekend or weekday, radio communications fill the air. Here are a few of the agencies that you can monitor: local police and fire, hospital, ambulance and medivac frequencies, civilian and military airplanes, coast guard, fish & wildlife, national parks, state parks, postal service, railroads, satellites, sports teams, and much, much more.

In the forthcoming chapters this book will provide you with thousands of frequencies to explore. You'll also discover just how easy it is to master the techniques that are necessary to fully enjoy the hobby of scanning.

In chapter 1, you'll learn how to select a scanner radio that will satisfy your expectations, without busting your budget. Chapter 2 provides you with easy to understand facts about indoor and outdoor scanning antennas. Coaxial cable and monitoring accessories are covered in chapter 3. In chapter 4, you'll learn about the rules and regulations that govern the radio spectrum. Best of all, you'll learn how to use the rules to your advantage. Chapter 5 is an up-to-the-minute report on the legal aspects of scanning. You'll learn about the new listening laws and how they will impact the hobby of scanning. In chapter 6, you'll learn exactly what's out there to hear—millions of conversations on the land, sea, and air. You will also learn that you can go beyond the sky and listen to radio communications from the Space Shuttle and orbiting satellites! Chapter 7 provides a comprehensive frequency listing in alphabetical order. If you want to monitor the marine boating frequencies, simply run your finger down the page to the "M" listings. Computers, and their role in radio communications, are explained in chapter 8. You'll also learn the secrets of monitoring computer controlled systems that were designed to prevent eavesdropping. How to design and build a professional listening post is covered in chapter 9. No matter where you live (single home, apartment, or condo) there are plenty of tips and ideas for everyone. Chapter 10 lists the various magazines, clubs, and organizations that are dedicated to the hobby of scanning. Chapter 11 provides answers to the most commonly asked questions.

Are you ready to begin? If so, strap yourself into a comfortable chair. We are about to begin our exploration of the mysterious, intriguing, and exciting hobby of scanning!

1
CHAPTER

Choosing a scanner

In today's high tech world, radio communications are playing a vital role in our every-day lives. In the Los Angeles area alone, there are approximately 60,000 handheld transmitters (walkie talkies) in use every day. From small business to top secret government agencies, radio communications are in use around the world. To hear the radio communications in your neighborhood, you'll need a scanner radio.

Choosing a scanner can be a confusing task. The electronic market contains an assortment of models with price tags that range between $100.00 and $1000.00. To help you choose wisely, here are a few basic guidelines to consider.

Frequency ranges

The common scanning frequency ranges are: 30.0 to 50.0 MHz (VHF Low Band), 150.00 to 174.00 (VHF High Band), and 450.0 to 512.0 MHz (UHF Band). These frequency ranges can be found in most moderately priced scanner radios ($100 to $150). Some models will also feature the civilian aircraft band, 118.0 to 137.0 MHz. Readers that live within a hundred miles of a commercial airport (Fig. 1-1) will be capable of hearing commercial airline pilots talking to control towers. It's also possible to hear ground and maintenance crews if you're between 25 and 50 miles from the airport.

Microwave mobile

Another popular frequency range is the new 806.0 to 960.0 MHz band. This band is utilized by nearly every large city in America. The band is often referred to as the "800-MHz band," or "Microwave Mobile Band." Cellular car phones, police, fire, and many other public and private agencies operate on these frequencies. Scanner radios that include the 800-MHz frequencies are more expensive than conventional models.

1-1
Commercial airports can be monitored on your scanner radio.

Military air

The military air frequencies can be found between 225.0 and 400.0 MHz. If you live within a hundred miles of a military air installation (Fig. 1-2), consider a scanner radio that can monitor the military air band. Fighter pilots, helicopter pilots, military control towers, and war games are just a few of the interesting conversations that can be monitored.

Continuous coverage

A scanner radio that can monitor the military air frequencies is usually referred to as a "continuous coverage" receiver (Fig. 1-3). In these models, the frequency ranges are not separated into specific bands. The manufacturer simply provides the low and high frequency limits. For example: Continuous coverage between 25 and 1300 MHz indicates that the radio can monitor all the frequencies between 25.00 and 1300.00 MHz. The frequency ranges for continuous coverage scanner radios are listed in Table 1-1.

1-2 Military aircraft bases are popular monitoring targets.

1-3 An example of a continuous coverage receiver.

**Table 1-1. Frequency ranges of
a continuous coverage receiver**

Frequencies
25.00 to 520.0
760.00 to 823.945
851.00 to 868.945 and
896 to 1300 MHz continuous.

Exceptions to the rules

As with all rules, there are exceptions. By law, the cellular phone frequencies (823.0 to 851.0 and 870.00 to 896.00 MHz) have been eliminated from general coverage scanner radios. There are, however, many scanner radios that were manufactured prior to the passage of the Telephone Disclosure and Dispute Resolution Act of April 26, 1994. Cellular capable scanner radios that fall into this category may be sold privately, but commercial sales will be prohibited after existing stocks are sold.

Look before you buy

As you have learned, the frequencies that a scanner radio can monitor are very important. Before you purchase a programmable scanner radio, ask to see the instruction booklet and study the frequency specification page (Fig. 1-4). If you're interested in listening to a particular frequency range that can't be found on the specification page, you'll need to consider another model.

Internal specifications

In addition to the frequency range of your scanner radio, you should also be interested in the radio's internal specifications. Generally speaking, the internal specifications can be limited to sensitivity, selectivity, intermodulation, and images. These terms are common to the hobby of scanning, and every scanner buff should become familiar with the following explanations.

1) Sensitivity: The weakest signal that the radio can capture. The lower the number, the better. Most scanner radios have a sensitivity of 0.5 microvolt.
2) Selectivity: The ability to reject adjacent frequency interference. The average selectivity for a scanner is 30 kHz.
3) Intermodulation: Strong signals that mix together and cause interference.
4) Images: Internally produced in all scanner radios. They are duplicates of the original signal. Images will be thoroughly discussed in a future chapter.

Nearly all scanner radios have good sensitivity and marginal selectivity. Most units suffer from intermodulation and are easily overloaded by strong signals (images). In a later chapter, you'll discover how to limit the effects of these and other internal problems.

Uniden Corp.

1-4
Check your instruction booklet for important information.

Base, mobile, and portable

In addition to selecting a particular model, consumers can choose between mobile, base, and portable, or handheld units. Mobile scanners can be installed in a vehicle and monitored while on the road. Base units are primarily used in the home or office. Handheld models are powered by batteries and can be carried to various locations.

There are specific units with multiple capabilities. For example: An ac wall transformer can power a handheld or 12-V mobile scanner radio from a base location, and certain handheld units can be temporarily installed in and powered from a vehicle.

At this point, you should be capable of choosing a scanner radio that fits your needs and budget. You'll be happy to know that the popular scanner radios from Radio Shack, Uniden, AOR, Yaesu, and Kenwood have excellent sensitivity and reasonable selectivity. Again, your main concern will probably focus on the specific frequency ranges of each model. Have fun and choose wisely.

Tabletop scanning

Nothing can compare to the thrill of bringing home your first scanner radio. As you open the box and remove the radio, the anticipation becomes nearly unbearable.

If you're a typical consumer, the anticipation of hearing your first scanner conversation will probably compel you to plug the radio into an outlet and completely disregard the instruction booklet. As we have already mentioned, you don't need a technical background to operate a scanner radio. But you should at least be familiar with the basic controls and terminology. The following information is provided as a supplement to the manufacturer's instruction booklet.

Power on

Turn on your scanner radio by rotating the off/on volume control clockwise. The unit will start scanning. Rotate the squelch control clockwise until you hear a loud hissing noise from the speaker. If you can't hear the noise, increase the volume. Slowly turn the squelch control clockwise until the noise stops. At this time, your scanner radio should start quietly scanning. It is important to note that the radio won't scan unless the squelch is set to eliminate the "hiss" that is heard between transmissions.

Scanning window

The window on your scanner radio is referred to as a liquid crystal display (LCD). The window will display the channel number, the frequency, and several additional functions. Figure 1-5 features the LCD of Radio Shack's PRO-2006. At first glance, the amount and variety of information that is displayed seems to be overwhelming. But in reality, you'll never see all of the information displayed at the same time. Each indicator represents a single specific function that you control via the keypad.

```
                         1  2  3  4  5  6  7  8  9  10
    Scan            Bank  -  -  -  -  -  -  -  -  -  -
    Manual
    Search          300ch     162 . 475  MHz
    Priority    P
    Program         Lock-out   Delay    AM NFM WFM    12 . 50 kHz
```

Radio Shack

1-5 The LCD window.

Basic keys

A typical scanner radio keypad can be seen in Fig. 1-6. Since keypad configurations and control panels differ among manufacturers and models, don't be alarmed if your scanner's keypad doesn't match the one shown. To become familiar with the actual controls on your scanner radio, look for the following:

Manual key Places the unit in the manual mode. Each time the button is pressed, the receiver will advance one channel.

Scan key The unit will automatically scan the entered frequencies.

Delay key Holds the receiver on an active channel briefly so that both sides of the radio conversation can be heard.

1-6 Keypad controls.

Priority key Establishes a priority channel for a special frequency that you choose (fire, police, etc.). When a call is received on the priority channel, the receiver automatically switches to the priority channel.

Lockout key Locks out channels and prevents them from activating the receiver.

Speed key Selects the scan speed. Scan speeds vary between models. The average scan speed is approximately 16 channels per second.

Enter key Enters the displayed frequency into an available memory.

Program key Prepares the internal computer chip to accept a specific frequency.

Monitor key Halts the search mode and allows a specific frequency to be monitored.

Clear key Clears/removes the frequency entered.

Number keys Numbered touch keys for entering frequencies.

Limit key Sets the high and low limits in the search range.

Arrow keys (^) Arrows change search direction to ascending or descending.

Mode key Sets mode to AM/NFM/WFM. (These various settings will be discussed in a future chapter.)

Step key Sets scanning increments to 5, 12.5, and 50 kHz.

Rear panel controls

It's also possible for the rear panel of your scanner radio to contain control switches. The rear panel on Radio Shack's PRO-2006 contains a sliding switch that adjusts the sensitivity to incoming signals. And although manipulation of the switch is rare, it merits consideration when the unit is permanently mounted. Figure 1-7 illustrates the rear panel of the PRO-2006 scanner radio. A detailed list and explanation of each component appear below:

Antenna connector You can utilize the indoor telescoping antenna supplied by the manufacturer, or use an outside antenna.

ac plug Connects to a standard ac wall outlet.

Memory backup Requires a 9-V battery to prevent loss of stored frequencies if ac power is interrupted.

13.8 Vdc jack Permits radio to be used with an external 12-V power source.

Extension speaker Used to connect an external speaker.

Tape-out jack For connecting a tape recorder to the radio.

Restart switch Resets and stabilizes an erratic LCD display.

ATT (dB) switch Increases or decreases the sensitivity of the radio to incoming radio signals. In metropolitan areas, interference from strong signals can be minimized by placing the switch in the "10-dB" position.

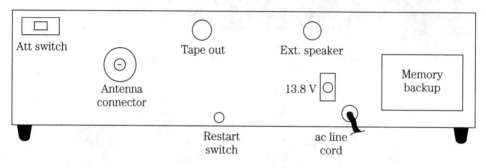

1-7 Rear panel controls and features.

Easy learning

Understanding and manipulating the controls on your scanner radio is not a difficult task. In most instances, all that's required is a little time, patience, and common sense. If you purchased a used scanner radio and don't have the manufacturer's instructions, the explanations provided in this chapter can be applied to practically any scanner radio on today's market. Best of all, you can't break your scanner radio by pushing the wrong button. So grab a cup of coffee and spend a few minutes familiarizing yourself with the controls on your scanner radio. I guarantee that it will be the start of an exciting and intriguing relationship.

2
CHAPTER

Antennas

The rooftop television antenna is probably the most common outdoor antenna in the world. Drive through any neighborhood and you'll see a variety of rooftop TV antennas (Fig. 2-1). As you examine the skyline in your neighborhood, you may also notice that certain rooftops are home to a very peculiar array of strange looking antennas. The classic "antenna farm" can be seen in Fig. 2-2. In this example there are several different types of antennas that seem to be propagating on the roof.

2-1 A variety of antennas can be seen on neighborhood rooftops.

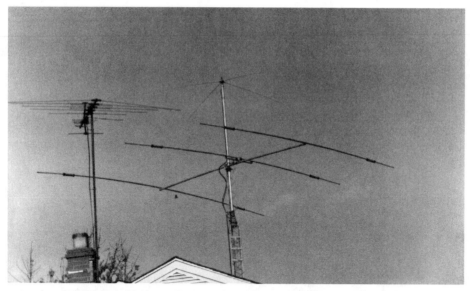

2-2 Radio hobbyists will often erect several different types of antennas.

In this chapter you will discover that there are many different types of antennas. Scanning antennas, like television antennas, come in a variety of shapes and sizes. Figure 2-3 shows a scanning antenna that could be mistaken for a form of abstract art.

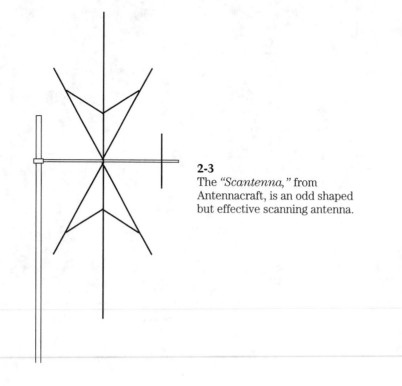

2-3
The *"Scantenna,"* from Antennacraft, is an odd shaped but effective scanning antenna.

At this moment, choosing the correct antenna for your particular scanning needs may seem like an impossible task. Should you simply use the small, factory antenna that was packaged with your scanner radio? Or would it be more practical to erect an outdoor antenna? The answers to these questions, and hundreds more, can be found in this chapter. We begin by discovering how antennas work.

How antennas work

All antennas are electrical conductors. The purpose of a scanning antenna is to receive a radio wave. The antenna converts the radio wave into electrical energy that is passed to the receiver. Since our antennas are usually located at considerable distance from the transmitter sites, the energy (signal strength) is measured in millivolts (thousandths of a volt). In most instances the signal strength of a received signal is less than a few microvolts (millionths of a volt).

Vertical vs. horizontal

As shown in Fig. 2-4, television antennas and FM radio antennas are mounted horizontal to the earth. Antennas that are mounted in this manner are referred to as "horizontally polarized" antennas. The antennas used with our scanner radios are "vertically polarized." For this reason, scanning antennas are erected vertically (Fig. 2-5). The two distinct antenna types help to prevent radio communications from police and fire departments, for example, from interfering with television reception.

Horizontal polarization for TV and FM antennas is standard in the United States. Since most man-made noise is vertically polarized, horizontal polarization was chosen because it helps to reduce interference from vehicle ignitions, neon signs, electric motors, hospital equipment, and thousands of other sources that are used in every neighborhood.

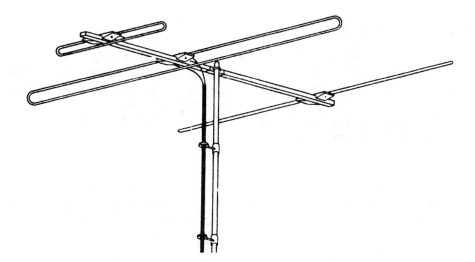

2-4 Television and FM radio station antennas are mounted horizontal to the earth.

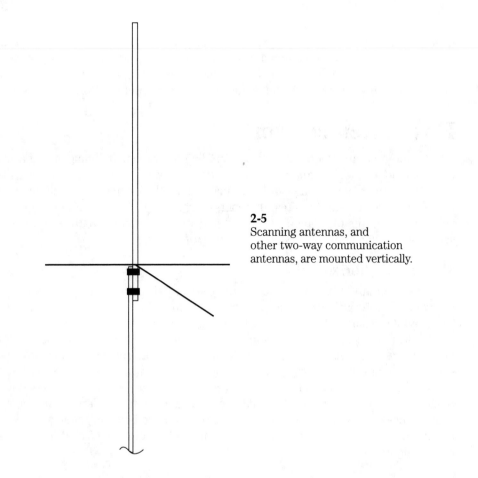

2-5
Scanning antennas, and
other two-way communication
antennas, are mounted vertically.

Radio waves

Radio waves are invisible and travel at the speed of light, 186,000 miles per second. Radio waves are considered to be a form of radiant energy, similar to heat or light. And as you may already know, radiant energy is affected by air, water, ground terrain, trees, and buildings. Radio waves can easily pass through wooden structures, but cement and metal structures can limit, or completely stop, radio waves from reaching our scanning antennas (Fig. 2-6).

Ground transmissions

The majority of radio waves that are received by our scanning antennas will be ground transmissions. Ground waves stay close to the earth and are also affected by natural and man-made objects. Figure 2-7 illustrates three popular mounting locations.

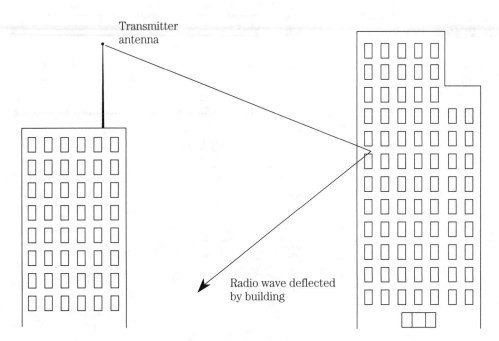

Transmitter
antenna

Radio wave deflected
by building

2-6 Thick concrete and metal surfaces can limit or completely stop radio waves from reaching an indoor antenna.

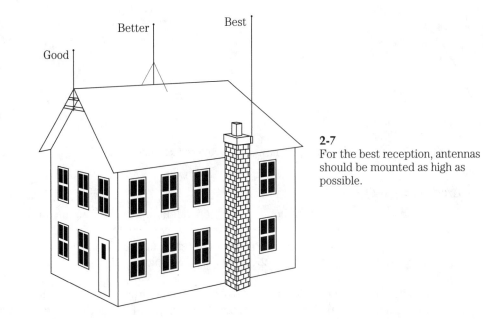

Good

Better

Best

2-7
For the best reception, antennas should be mounted as high as possible.

Skip

Oxygen, hydrogen, helium, and nitrogen make up the atmosphere that surrounds the earth. Ultraviolet radiation from the sun ionizes the gases at the higher altitudes and layers form around the earth. The ionized layers of gas have the ability to bend and reflect radio waves back to the earth (Fig. 2-8).

Radio waves radiate (travel) in all directions. Waves that radiate toward the upper atmosphere are sometimes reflected and return to earth. The reflected wave is commonly called "skip." And, as shown in Fig. 2-9, the distance traveled by the radio wave can be considerable. Skip is more likely to occur at radio frequencies below 60 MHz. Since skip is dependent upon the sun's radiation to ionize the atmosphere, it is more likely to occur during daylight hours.

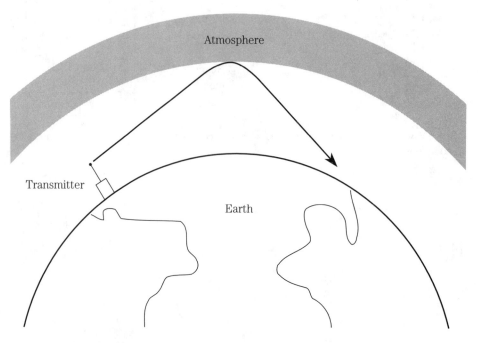

2-8 The atmosphere has the ability to bend and reflect radio waves.

Distance traveled

The strength of the received radio signal is dependent upon the amount of power at the transmitter. Other factors include antenna gain, distance from the transmitter to the receiver, the location and type of antenna utilized, and the path that the radio signal travels. If the path is obstructed by buildings or hilly terrain, the distance traveled may be reduced.

Under ideal conditions, the radio waves that we will be monitoring on our scanner radios rarely travel more than a hundred miles. Sure, skip can travel thousands of miles, but skip is a rare phenomenon that occurs only sporadically. And since we won't be operating our scanner radios under ideal conditions (ground terrain, trees,

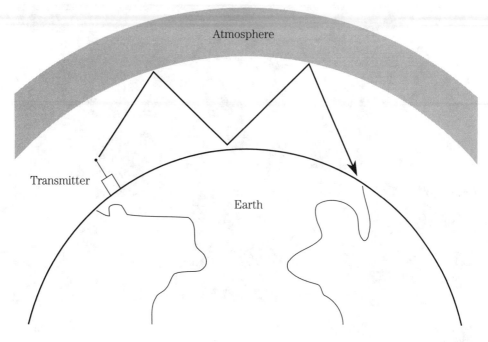

2-9 Radio waves that skip between the earth and atmosphere can travel thousands of miles.

and buildings will get in the way), it's more likely that our scanning range will be limited to about 60 miles.

If the 60-mile range seems limited, you couldn't be more wrong. The number of transmitters located within a 60-mile radius of your location can number in the thousands. To gain a better perspective of your monitoring area, we suggest using a road map. Refer to the map's scale and draw a 60-mile radius around your location. Finally, draw another radius that represents 100 miles. Your map should resemble the one in Fig. 2-10.

Outdoor antennas

With an outside antenna, it will be possible to monitor every city and/or town within the 60-mile radius. The second (100-mile) radius represents the area that will be more difficult to monitor. Your success in the 100-mile range will depend upon your location, antenna height, and both natural and man-made obstructions. Scanner buffs living on a hill, or on the top floor of an apartment tower, may discover that they can monitor signals from 100 miles away without difficulty.

Indoor antennas

If you're using the small indoor antenna that came with your scanner radio your monitoring range will be limited to local reception, approximately 25 miles. Again, don't jump to any hasty decisions. Readers living within a large city will probably discover that a small, indoor antenna can fully satisfy their monitoring needs. Readers living in more rural locations can extend their listening range with an outside antenna.

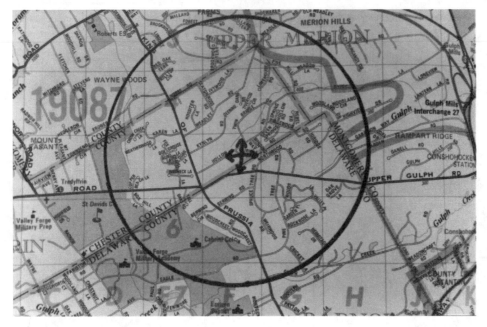

2-10 Drawing a circle around your monitoring area helps to establish your listening boundaries.

Types of antennas

Ground plane

The ground plane antenna is probably one the most popular antenna designs. During the CB radio era ground plane antennas, similar to the one featured in Fig. 2-11, could be seen in neighborhoods throughout the United States.

A ground plane antenna usually has three elements that extend from the base. The length and number of elements that extend vertically will vary between models. One of the key features of the ground plane is the ability to receive radio signals from all directions, that is, it is omni-directional. Since the antenna does not have a front or back, it can be installed without the need to aim or point the antenna in a specific direction.

Ground plane antennas are often advertised as having the ability to exhibit gain. The measurement of gain is expressed in decibels (dB). Gain is another way of saying that the antenna has the ability to amplify or boost the received radio signal. If an antenna can boost the received signal by 3 dB, that represents an increase of approximately 50 percent. Increasing the strength of the radio signal in this manner will provide the listener with a clearer signal and will extend the listening range.

Beam antenna

Any antenna that provides high gain will also be highly directional. Directional antennas are commonly called "beam antennas." The Grove Scanner Beam can be seen in Fig. 2-12. This particular type of antenna has a front and a back, and must be

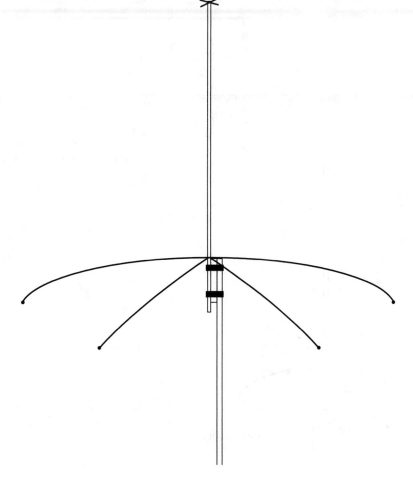

2-11 Citizen's Band ground plane antennas were common during the '70s.

pointed in the direction of the transmitter. Signals arriving at the back or sides of the beam will be reduced in strength, while signals at the front are amplified.

To receive signals from various locations, beam antennas are usually rotated with an electric motor that is connected to the mast. A common TV antenna rotor (Fig. 2-13) can be used to turn a lightweight beam antenna for strongest signal.

Discone antenna

The discone antenna is shown in Fig. 2-14. Discone antennas are omni-directional, but they do not exhibit gain. Discone antennas are often advertised as having a 25.00- to 1000.00-MHz response. In reality, discone antennas have a very narrow field of performance. They usually provide good reception between 100.00 and 300.00 MHz. Reception on the upper or lower side of this threshold will be compromised. For improved reception of signals below 100 MHz or above 300 MHz, a ground plane or beam antenna are the recommended choices.

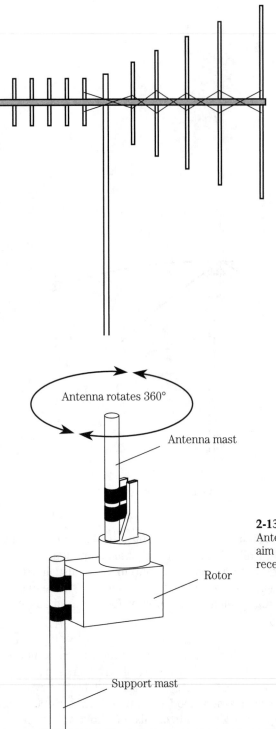

2-12
The Grove *Scanner Beam* is a directional antenna that must be pointed in the direction of the transmitted signal. <small>Grove Enterprises</small>

Antenna rotates 360°

Antenna mast

2-13
Antenna rotors can be used to aim a beam antenna for the best reception.

Rotor

Support mast

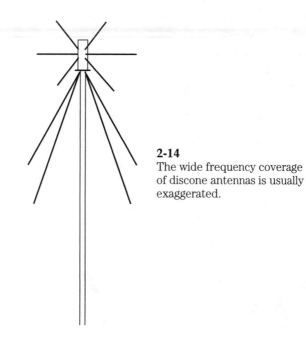

2-14
The wide frequency coverage
of discone antennas is usually
exaggerated.

Longwire antenna

A longwire antenna is nothing more than a random length of wire. In Fig. 2-15, a longwire antenna is shown hanging from a tree branch. The antenna is connected to a length of coaxial cable (you'll learn about coaxial cable in chapter 3). The coax passes through the structure (Fig. 2-16) and connects to the back of your scanner radio.

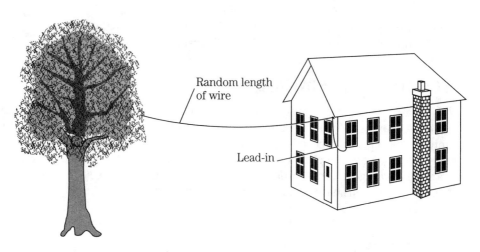

Random length
of wire

Lead-in

2-15 A random length of wire can be used as an effective antenna.

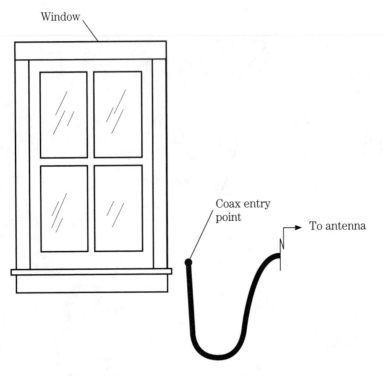

2-16 A typical entry point for coax cable is near a window.

Mobile antennas

Mobile scanning antennas also come in a variety of sizes and shapes (Fig. 2-17). There are models designed to receive one specific frequency and models that can receive a broad range of frequencies between 30 and 1000 MHz. Mobile scanning antennas exhibit gain and unless they are made for a specialized application, such as direction finding, they receive signals from all directions.

Frequency response

No matter what type of antenna that you choose, it's a good idea to match the frequency response of your antenna with your scanner radio. For example: if your scanner radio cannot receive beyond 512.00 MHz, it makes little sense to purchase a wide coverage antenna that can monitor through 1000.00 MHz. Purchasing a wide coverage antenna from the start, however, may not be such a bad idea. If you eventually purchase a wide coverage receiver another trip to the roof won't be necessary!

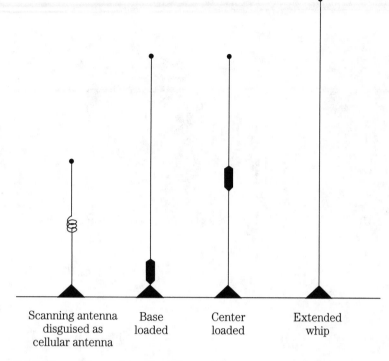

Scanning antenna   Base   Center   Extended
disguised as   loaded   loaded   whip
cellular antenna

2-17 There are a variety of mobile scanning antennas on today's market.

Antenna installation

Your scanning antenna's performance will be dependent upon several key factors. Objects that surround an antenna may affect its performance. We have already learned that buildings and trees are common obstacles that can adversely affect antenna performance. The electrical wires that supply your home with power also affect antenna performance. Electrical wires are common sources of other bothersome noises that can be received by your scanning antenna.

Mounting your antenna

Ideally, your antenna should be mounted above nearby obstacles. Figure 2-18 illustrates the "line-of-sight" rule that is used to guarantee optimum results. Mounting your antenna so that it can "see" over existing structures, however, may not be practical. A typical example can be seen in Fig. 2-19. In this situation, it would be nearly impossible to raise the antenna above nearby buildings. The alternative is to mount the antenna in another location or to simply mount it at a reasonable distance above the roof (Fig. 2-20), and hope for the best.

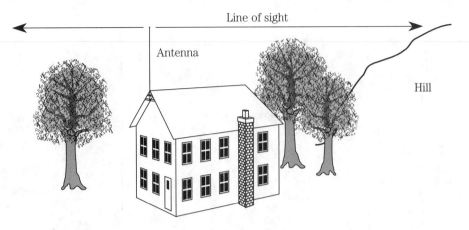

2-18 The line-of-sight rule is a guideline for determining antenna height.

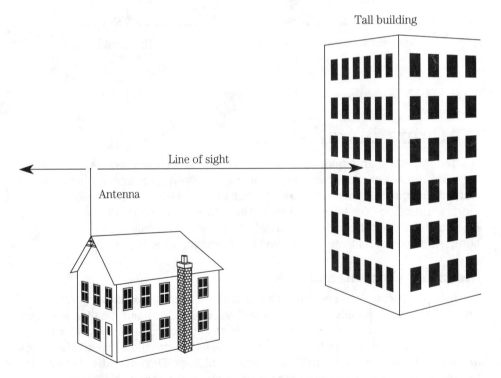

2-19 Mounting your antenna above the line of sight might not be practical.

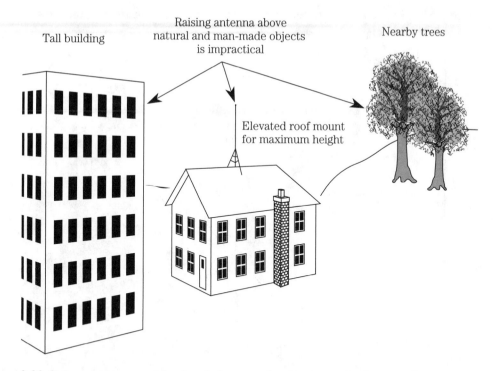

Tall building

Raising antenna above
natural and man-made objects
is impractical

Nearby trees

Elevated roof mount
for maximum height

2-20 If the antenna cannot be placed above nearby structures, simply mount it at a reasonable distance above the roof.

Safety

Serious injuries and deaths have resulted during the installation of outdoor antennas. People fall from roofs or they are electrocuted by nearby power lines. To prevent accidents, the distance to any power line should be at least twice the length of the antenna, the mast included (Fig. 2-21). Metal ladders should be avoided. When a metal ladder contacts a power line, it becomes an extension of that line and can cause a fatal shock.

Attic antennas

Readers who are afraid of heights, or who are not comfortable with the idea of climbing onto their rooftops, should consider placing an antenna in the attic. Attic mounted antennas (Fig. 2-22) perform very well and may even rival the performance of an antenna mounted outside. Before you venture out on the roof, place your scanning antenna in the attic and scan your favorite frequencies. If the reception is satisfactory, don't hesitate to permanently install the antenna in the attic. The installation is safer and faster, and your antenna isn't subjected to weather related maladies.

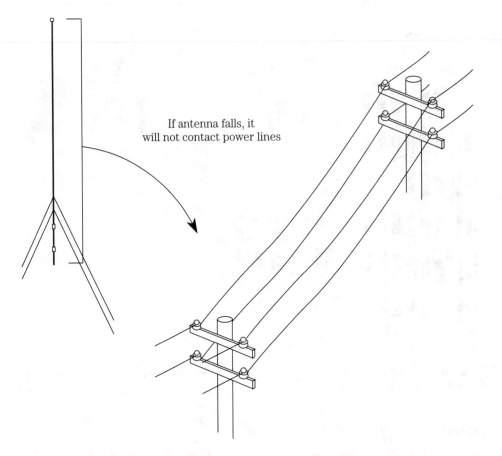

If antenna falls, it
will not contact power lines

2-21 The distance between power lines and your antenna should be at least twice the length of the antenna and mast.

Finally, if you don't have the experience but simply can't live without an outside antenna, contact a professional antenna installer. Check the yellow pages under "Antennas" or "Television."

Checklist

A do-it-yourself antenna installation should be carefully planned. Here is a checklist that you should review before beginning the installation.
1. Don't do your installation during windy or inclement weather.
2. Never work alone. Have a helper on the ground, ready to respond to an emergency.
3. Assemble the antenna on the ground and use a rope to hoist it to the roof.
4. Plan ahead and make a tool list. Place your tools in a canvas bag or similar container, and hoist them to the roof with a rope.
5. Check your local building codes. Your particular town or city may have special requirements for erecting an outside antenna.

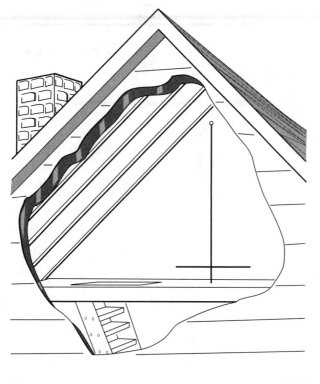

2-22
Attic mounted antennas are
easy to install.

Lightning

During an electrical storm, any rooftop antenna is at risk from lightning. If the antenna is not properly protected, lightning damage to your equipment and home can be severe. When lightning strikes an antenna, the electrical current begins to seek the ground (earth). In Fig. 2-23, two possible paths are shown: 1) down the mast, and 2) down the lead-in (coaxial cable). In most instances the current will attempt to reach the ground by traveling down both paths.

Grounding

To prevent structural damage to your home, a heavy gauge wire must be connected between the antenna mast and the ground (Fig. 2-24). The ground wire must be securely connected to the mast and routed to ground by the shortest path. Sharp bends or kinks in the wire (Fig. 2-25) are not permitted. Bends and kinks may cause the electrical current to leave the wire and damage your home.

Ground wire

The ground wire must also be insulated from the structure with insulated standoffs (Fig. 2-26). Again, the ground wire must not touch any part of the structure. At the ground level, the grounding wire must be securely attached to a metal device that has been driven or buried in the earth.

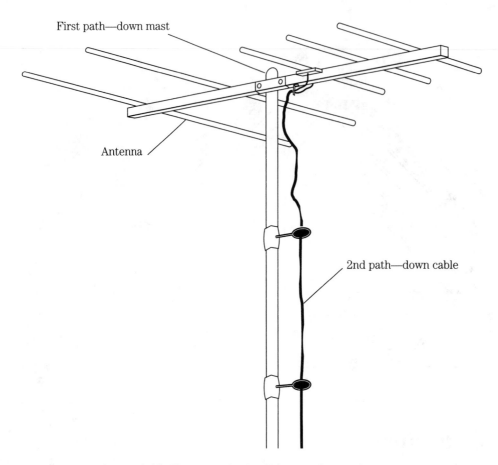

First path—down mast

Antenna

2nd path—down cable

2-23 The two paths that lightning can travel.

Grounding requirements

Because antenna grounding requirements vary between communities, we have purposely omitted any reference to wire and grounding rod specifications. Prior to erecting or grounding an antenna, readers are strongly urged to follow the guidelines of their local building codes.

Lightning arrester

As we have already mentioned, the second path that lightning can take is down the coax. This results in extensive damage to the equipment connected to the antenna. To avoid this damage, the coaxial cable should contain a lightning arrester (Fig. 2-27). A separate ground wire is attached to the arrester and to a ground rod.

During a lightning strike, the arrester directs the surge of electrical current to the ground rod, preventing damage to equipment in the home. Lightning arresters will also prevent your antenna from collecting static charges from the atmosphere and nearby lightning strikes. Again, don't install a lightning arrester without consulting your local building codes. There are several types of arresters on the market and your community might require the use of a specific make or model.

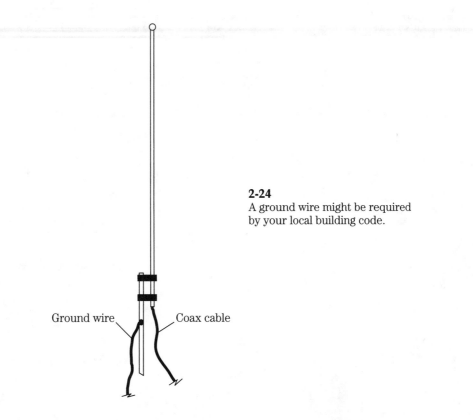

2-24
A ground wire might be required
by your local building code.

Ground wire   Coax cable

Antenna mounts

Towers

The antenna tower is shown in Fig. 2-28. Towers are usually made of sections
that are bolted together. To support the tower, a thick concrete base is installed in
the ground. In rural areas, towers are used to raise television antennas to heights
that cannot be reached with a standard roof mount. Amateur radio operators (hams)
also use towers to support heavy or bulky antennas that cannot be roof mounted.

Roof mounts

There are a variety of antenna mounts that can be installed on a roof. A few of the
more common mounts are shown in Fig. 2-29. An A-frame mount is shown in Fig. 2-29A.
This mount is very popular because it is not directly attached to the roof. The wall
mount in Fig. 2-29B can be used on cinder block, cement, or brick buildings. The chimney mount in Fig. 2-29C is another roof mount that does not disturb the roof surface.

Guy wires

Roof mounts that are directly attached to the roof are shown in Fig. 2-30. The
mount shown in Fig. 2-30A can be installed over the roof peak or installed on a flat
roof as shown in Fig. 2-30B. When this type of mount is used, guy wires are required

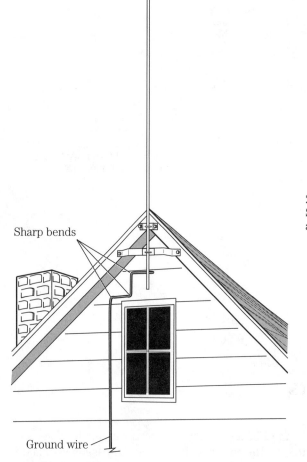

Sharp bends

Ground wire

2-25
Sharp bends in the ground wire
are not permitted.

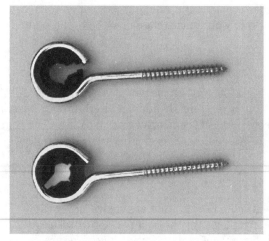

2-26
Insulated stand-offs prevent the
ground wire from contacting the
building.

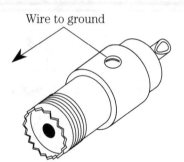

Wire to ground

2-27
Lightning arresters can prevent serious damage to your equipment.

2-28
Antenna towers are used to provide additional height and support for single and multiple antennas.

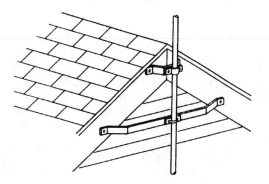

(A) A frame mount.

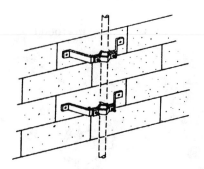

(B) Wall mount.

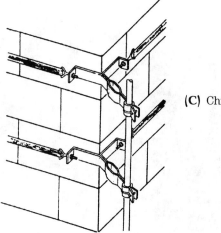

(C) Chimney mount.

2-29 There is an antenna mount for nearly every application.

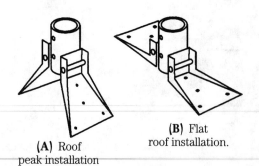

(A) Roof
peak installation

(B) Flat
roof installation.

2-30
Roof mounts can be attached
directly to the structure.

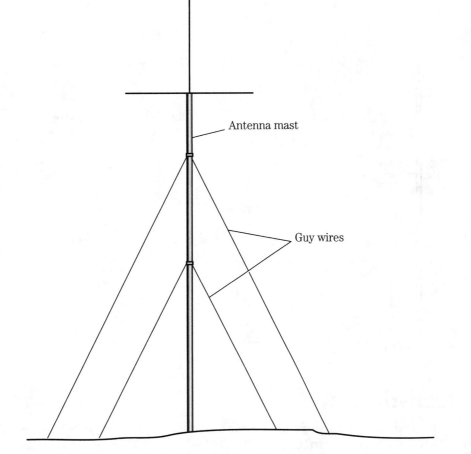

Antenna mast

Guy wires

2-31 Guy wires help to support the antenna and prevent swaying.

to support the mast (Fig. 2-31). Guy wires are made from galvanized wire that is fastened directly to the antenna mast. The wires are tightened by turnbuckles that are attached to each wire. The guy wires should be insulated from the roof by using insulated anchors or by placing an insulator in each wire (Fig. 2-32).

Prior to tightening the guy wires, make sure that the mast is vertical and that the wires are tightened equally. Do not over-tighten the guy wires. Moderate tension to prevent the mast from swaying is all that's required. Antenna mounting hardware, including guy wire, is available from Radio Shack stores.

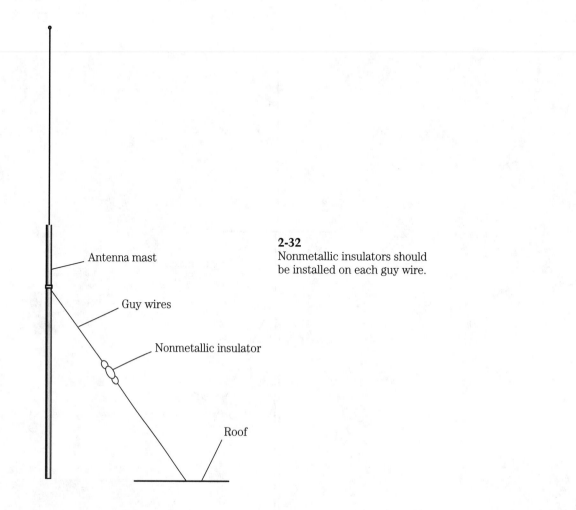

Antenna mast

Guy wires

Nonmetallic insulator

Roof

2-32
Nonmetallic insulators should
be installed on each guy wire.

Realistic goals

In this chapter, the goal was to provide the reader with a basic understanding of how antennas work and perform and to familarize the reader with antenna mounting procedures and safety precautions. In the real world, antenna design and theory are far more complicated. For additional information on antennas, contact your local library or amateur radio (ham) store.

3
CHAPTER

Coaxial cable

The wire that connects the antenna to your scanner radio is called *coaxial cable*, or coax. Cable television companies use coax to connect your television to pay TV broadcasts (Fig. 3-1). Coax is also used to connect your AM/FM car radio to the antenna mounted on the fender.

3-1 Coaxial cable is widely used in cable television installations.

33

The physical characteristics of coaxial cable can be seen in Fig. 3-2. Coax is available in a variety of sizes (Fig. 3-3), but the basic design does not change. Some coaxial cables have stranded wire center conductors, others utilize a solid wire. The dielectric material also varies in size and material. The outer conductor (shield) might be single strands of wire or multiple strands braided together. Finally, an outer vinyl jacket provides protection from moisture and other contaminants.

The physical differences of each cable are directly related to how the cable will be used. The larger diameter cable usually offers low loss with high power capabilities. The larger diameter of RG-8, for example, permits the cable to handle the higher power from ham radios and other transmitters. The smaller diameter of RG-8x mini cable compromises power handling ability for flexibility and ease of concealment.

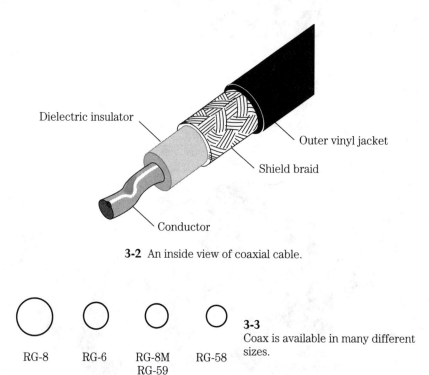

Dielectric insulator

Outer vinyl jacket

Shield braid

Conductor

3-2 An inside view of coaxial cable.

RG-8   RG-6   RG-8M   RG-58
                    RG-59

3-3
Coax is available in many different sizes.

Receiving a strong signal

Because you will not be transmitting through your coax cable, there's no reason to be concerned with power ratings. Your primary concern should be focused on how much of the received signal is lost per foot. In Table 3-1, you can easily see that RG-6 offers the lowest loss per foot. An explanation of the negative numerical values can be found in Table 3-2. By comparing both tables, you can see that at 450 MHz, RG-6 will lose 4.4 dB, or 60 percent of the signal strength. Although the loss is significant, it is lower than the remaining cables. The low loss, small diameter, and flexibility of RG-6 make it ideally suited for our listening hobby.

Table 3-1. Coaxial cable loss per foot

Frequency	50.00	150.00	450.00	1000.00 MHz
RG type—6=	−1.4	−2.4	−4.4	−8.3 dB loss
8/M	−2.3	−4.1	−8.8	−17.1
8	−1.4	−2.4	−4.6	−8.6
58	−3.9	−5.9	−10.2	−17.0
59	−2.5	−4.2	−7.3	−15.0

Table 3-2. Loss comparison chart

dB loss	Percentage of signal loss
−1	21%
−2	37%
−3	50%
−4	60%
−5	69%
−6	75%
−7	80%
−8	84%
−9	87%
−10	90%

Impedance

One common argument against the selection of RG-6 cable is referred to as *impedance*. The electrical characteristics of any length of coax is determined by the cable's resistance, capacitance, and inductive properties. Maximum signal transfer from an antenna to the receiver will occur when the electrical properties (impedance) of the antenna match the electrical properties (impedance) of the coaxial cable. If the system is mismatched, the impedance of the antenna and/or coax must be corrected.

If it sounds confusing, don't be concerned. In the hobby of scanning it is nearly impossible to maintain a constant impedance match. The wide frequency range of the equipment simply won't allow a perfect match between the antenna and cable, so concentrate on using low-loss, inexpensive cable that can be easily installed. RG-6 is the ideal candidate.

Coax cable splicing

Coax cable must be installed in one continuous length. It cannot be spliced like electrical wire (Fig. 3-4). Any type of splice will drastically reduce the cable's ability to deliver a strong signal to your receiver. Excessive cable length can also reduce signal strength. An overall length of up to 100 feet is acceptable (Fig. 3-5). Signal loss rises sharply when cable lengths exceed 100 feet.

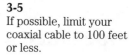

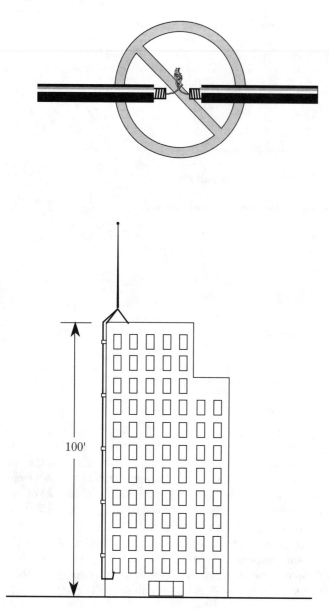

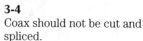

3-4
Coax should not be cut and spliced.

3-5
If possible, limit your coaxial cable to 100 feet or less.

Weather exposure problems

Coax cable that is exposed to the weather should be checked yearly for cracks and abrasions. Cable that has been contaminated by moisture must be replaced. To prevent moisture from entering the cable at the antenna connector, use a waterproof sealant (Fig. 3-6). Radio Shack's moldable sealant tape, Catalog #278–1645, is an ideal choice.

3-6 Use a waterproof sealant to protect coax connections.

Cable entry hints

There are several different methods that can be used to get the coaxial cable (feed line) to your scanner radio. Many hobbyists will replace the glass pane in a window with clear plexiglass. The plastic can be drilled to accept one or several feed lines (Fig. 3-7).

Another approach is to insert a wood panel beneath the window (Fig. 3-8). Like plexiglass, the wood can be drilled to accept a variety of cables. To deter theft the window should be secured with some sort of locking device.

Entrance points

The entry point of all feed lines should contain a drip loop (Fig. 3-9). The loop prevents water from traveling down the feed line and entering the structure. Do not kink or sharply bend the feed line. Simply allow the cable to make a natural loop that hangs below the entry point. To further protect the structure, a waterproof sealant can be applied to the point of entry (Fig. 3-10).

Factory installed connectors

If you're using coax cable with factory installed connectors the entry hole will need to accommodate the larger diameter of the connector (Fig. 3-11). After the cable passes through the opening, the excessive hole size must be plugged or sealed. Experienced hobbyists prefer to attach the connectors after the cable is installed. This prevents the drilling of entry holes that are excessively large.

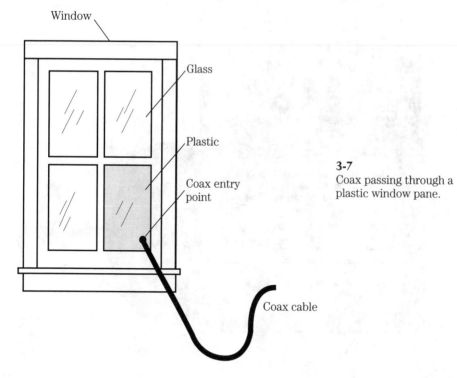

Window

Glass

Plastic

Coax entry
point

3-7
Coax passing through a
plastic window pane.

Coax cable

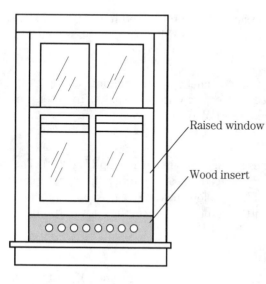

Raised window

Wood insert

3-8
Wood inserts can be drilled to
accept various coaxial cables.

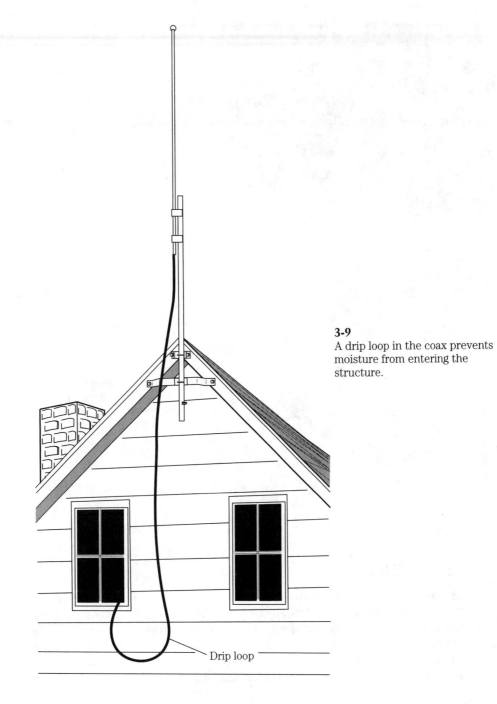

3-9
A drip loop in the coax prevents
moisture from entering the
structure.

Drip loop

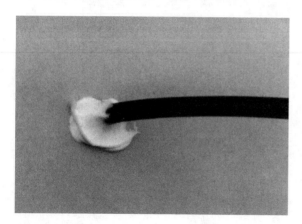

3-10
All entry points should be
sealed to prevent moisture
and insects from entering.

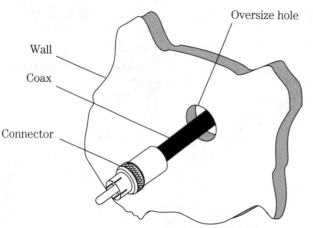

Oversize hole

Wall

Coax

Connector

3-11
Factory installed
connectors will require
large entry holes that
must be sealed.

Cable connectors

The antenna connections on the rear of your scanner radio and antenna will
vary. An assortment of connectors are shown in Fig. 3-12. Connecting your antenna,
coax, and scanner radio can be a confusing task. The connection on the rear of your
scanner radio will differ from the connection on your antenna. And if you purchased
coax with factory-installed connectors, it's probably safe to assume that one or both
of those connectors will fail to mate with your equipment.

Linking mismatched connectors

The first step to linking mismatched connectors is to identify the connectors in
your system. The most widely used connector is the PL-259 (Fig. 3-13). It is in-

3-12 There are dozens of coax connectors.

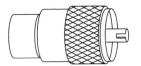

3-13
The PL-259 connector is widely
used in the hobby of scanning.

tended for heavier coax, like RG-8, but reducers (Fig. 3-14) are available for smaller diameter coax, such as RG-6. The "BNC" connector (Fig. 3-15) features low loss and is commonly used on present-day scanning equipment. The familiar "F" connector (Fig. 3-16) is widely used by cable TV companies and VCR manufacturers. Cable customers and VCR owners are probably accustomed to seeing "F" connectors on the cables that connect their equipment.

Developed as an improved version of the PL-259, "N" type connectors (Fig. 3-17) are primarily used in commercial satellite ground stations. However, "N" connectors are occasionally used on scanning equipment.

Installation hints

Connectors must be installed by someone who is familiar with basic hand tools. Skill with a soldering iron is another talent that may also be needed. There are solderless connectors on the market, but the majority of connectors require a combi-

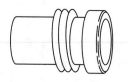

3-14
Reducers are used to match different sized coaxial cables with a variety of connectors.

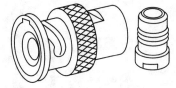

3-15
The low-loss BNC connector has become an industry standard on present-day scanning equipment.

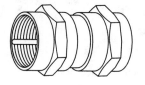

3-16
The F connector is common to cable television installations.

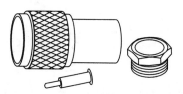

3-17
The N type connector, normally used by the military, is occasionally used on scanning equipment.

nation of stripping, crimping, and soldering. Figure 3-18 illustrates the required steps for installing a PL-259 connector. Follow the step-by-step instructions exactly as illustrated and don't forget to use acid-free solder. Installing "F" connectors on RG-6 cable can be accomplished without soldering. Figure 3-19 shows the necessary installation steps. The solderless "BNC" connector is probably the easiest to install. Figure 3-20 provides the necessary steps to follow.

It would take a separate book to illustrate the fastening procedures for every connector that is on the market. The three connectors mentioned above are the primary connectors that are encountered by the scanning hobbyist. If it becomes necessary to mate a pair of mismatched connectors, don't panic. The problem can probably be solved with an adapter.

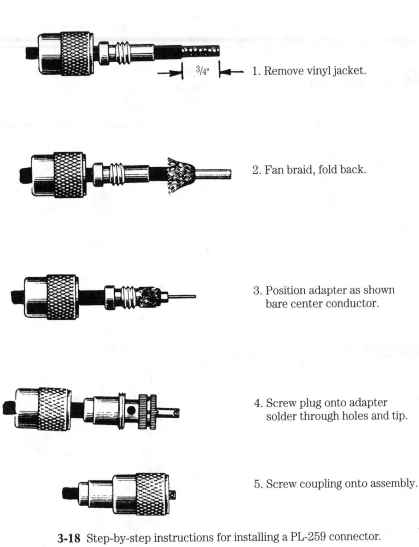

1. Remove vinyl jacket.

2. Fan braid, fold back.

3. Position adapter as shown bare center conductor.

4. Screw plug onto adapter solder through holes and tip.

5. Screw coupling onto assembly.

3-18 Step-by-step instructions for installing a PL-259 connector.

Adapters

Readers who do not have the skills to custom install connectors, will be happy to learn that a wide variety of adapters are available. Figure 3-21 shows a "PL-259 to BNC" adapter. This particular adapter will allow a PL-259 connector to mate with a BNC connector. Adapters are available for nearly every conceivable mismatch that the hobbyist may encounter. Adapters cannot, however, be used without caution. Adapters can cause a loss of signal strength. Using more than two adapters on a single coax cable (Fig. 3-22) can reduce the signal strength by more than fifty percent. Coaxial cable, connectors, and adapters are available from Radio Shack and amateur radio stores.

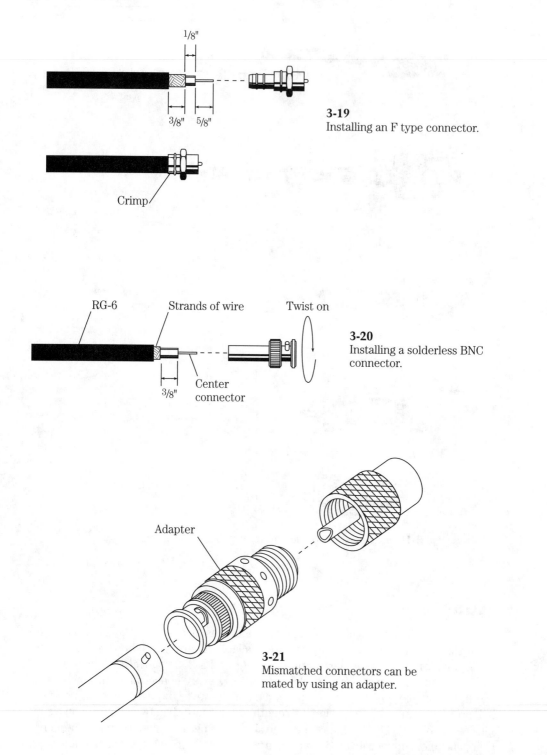

1/8"

3/8" 5/8"

3-19
Installing an F type connector.

Crimp

RG-6   Strands of wire   Twist on

3/8"   Center connector

3-20
Installing a solderless BNC connector.

Adapter

3-21
Mismatched connectors can be mated by using an adapter.

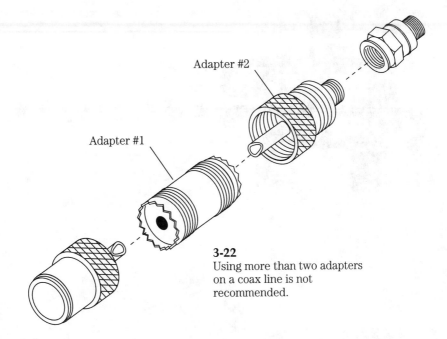

Adapter #2

Adapter #1

3-22
Using more than two adapters
on a coax line is not
recommended.

Accessories

Cable television accessories are commonly utilized in the hobby of scanning. The most popular cable TV accessory is the cable *splitter* (Fig. 3-23). A splitter is used when it's necessary to connect more than one device to a single cable (Fig. 3-24). The operating range of cable TV splitters and other accessories is usually marked on the component or listed on the packaging. The operating range is important because it limits the frequency ranges that we can explore. If the frequency range of a splitter is 70 to 500 MHz, it would be impossible to monitor a radio signal that was below 70 or above 500 MHz. Listeners who desire to explore beyond the listed operating range must remove the splitter or purchase a unit that includes the desired frequency range.

Signal amplifiers

Signal amplifiers, commonly called *pre-amps*, are used to increase the strength of weak signals (Fig. 3-25). It is important to remember that pre-amps amplify **everything** that is received by the antenna, including noise. Weak signals, as well as strong signals, are amplified and sent to the receiver. In the hobby of scanning, this usually produces a lousy radio signal that overloads the radio. Don't fool yourself into thinking that a pre-amp will extend or improve your listening range. Pre-amps are generally used by experienced hobbyists who are attempting to monitor a specific group of frequencies. When used for wide band reception, pre-amps usually fail to provide satisfactory results.

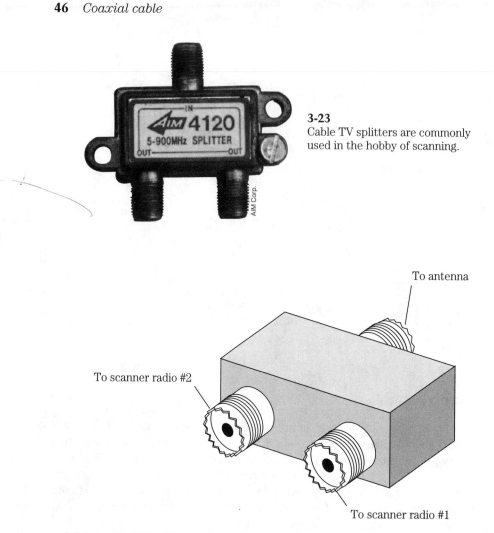

3-23
Cable TV splitters are commonly used in the hobby of scanning.

To antenna

To scanner radio #2

To scanner radio #1

3-24 A cable TV splitter can be used to connect two scanner radios to a single antenna.

BROADBAND RF PREAMPLIFIER
10 MHZ – 1 GHZ

IN   POWER   OUT

Ramsey Electronics

3-25
Pre-amps are used to increase the strength of weak signals.

Filters & attenuators

Annoying signals that interfere with normal reception can be controlled by utilizing several methods. The most popular solution is to install a *filter*. Suppose for a moment that you live near a television broadcast station. To prevent television signals from overloading your scanner radio, a special filter could be installed in the coax line (Fig. 3-26). The filter would prevent the television frequencies from reaching your receiver, but the scanning frequencies would not be blocked.

The second popular remedy is to install an *attenuator* (Fig. 3-27). These small, inexpensive devices reduce the strength of bothersome radio signals. Attenuators are rated according to the amount of signal that is reduced—a 3 dB rating is usually sufficient. Attenuators and filters are installed in the coax line with standard "F" connectors. Attenuators are available in Radio Shack and department stores. Scanning filters are specialty items that cannot be found on store shelves. To receive a free catalog that includes filters and other scanning accessories, contact Grove Enterprises, P.O. Box 98, Brasstown, NC 28902, or one of the other suppliers listed in chapter 10.

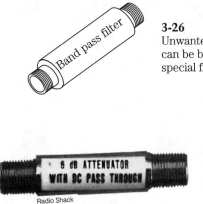

3-26
Unwanted radio signals
can be blocked by using a
special filter.

3-27
The attenuator is used to
prevent strong, nearby signals
from overloading your receiver.

Antenna switches

Antenna switches are used to connect your receiver to a variety of antennas. In Fig. 3-28, three antennas are connected to a three-position push-button switch. Any antenna in the group can be selected by merely pushing the corresponding button. The switch can also be installed in reverse (Fig. 3-29). When used in this manner, the switch can route the signal from a single antenna to three separate receivers.

Accessory cables

After-market items are connected between your antenna and scanner radio with short lengths of coaxial cable (Fig. 3-30). Cable of various lengths, with factory installed connectors, are available in the video department of Radio Shack and department stores. Don't buy the more expensive cables that feature gold plated connectors, your ears can't detect the difference.

3-28
Selection and/or isolation of
a particular antenna can be
accomplished by using an
antenna switch.

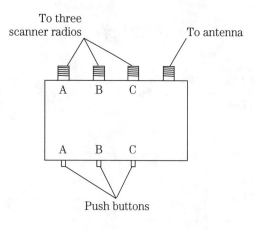

3-29
Antenna switches can be
installed in a variety of
positions.

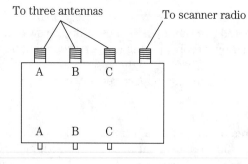

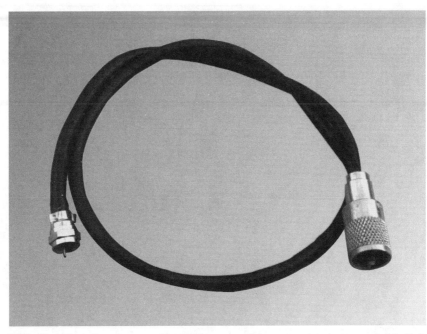

3-30 Short lengths of coax cable, with appropriate connectors, are needed to link accessories.

Connecting everything together

Antennas, cables, connectors, and accessory items. Could you put a complete system together? At first glance, it may seem like a monumental task. But if you tackle each component separately, it's fairly easy. If you experience difficulty, don't let your personal pride get in the way. Carry your antenna and scanner radio into the nearest Radio Shack store and ask the sales person to match them up with the proper connectors and/or adapters.

4
CHAPTER

Rules and guidelines

The Federal Communications Commission (FCC) is responsible for governing the radio spectrum. It is unlawful for any agency or individual to transmit on a frequency that has not been assigned for their particular use. A typical example is the FM radio broadcast band, 88.0 to 108.0 MHz. When you tune in your favorite FM radio station, you don't want to hear communications from CB radios or television stations. On the other hand, television viewers don't want their favorite TV show interrupted by FM radio transmissions.

Aircraft safety

The need for separate frequency ranges is especially apparent in the aircraft band. Airline pilots and air traffic controllers must have uninterrupted communications. The safety of airline passengers would be in jeopardy if the FCC allowed another agency to use the aircraft frequencies. Your local police, fire, and ambulance frequencies are as equally important. The fire department frequencies are separate from the police frequencies. A police officer in high speed pursuit doesn't want his radio communications interrupted by a fire truck responding to a trash fire.

Finding fingerprints

The strict, definitive guidelines that separate each agency can also provide the hobbyist with a *fingerprint* that can be used to identify individual frequencies. In Table 4-1, you will see that the radio spectrum contains numerous agencies that are assigned specific frequencies. Suppose that you're listening to the frequency of 418.90 MHz. The spectrum chart indicates that the frequency is used by the federal government. If you're listening to a conversation on 420.625 MHz, the spectrum chart will help you to determine that the frequency belongs to the amateur radio service.

Table 4-1. The radio spectrum

Frequency	User
30–50	Civilian & Govt. low band
50–54	Ham band (6 meters)
54–72	TV channels 2–4
72–76	Paging, repeater links
76–78	TV channels 5–6
88–108	FM broadcasts
108–136	Civilian aircraft VHF
136–138	Weather satellites
138–144	Military aircraft VHF
144–148	Ham band (2 meters)
148–150.8	Military bases
150.8–174	Civilian & Govt. high band
174–216	TV channels 7–13
216–220	Inland waterway navigation
220–225	Land mobile/ham (shared)
225–400	Military aircraft (UHF)
400–406	Weather balloons, telemetry
406–420	Federal Govt. land mobile
420–450	Ham band
450–512	Civilian UHF band
512–806	TV channels UHF
806–960	Cellular, microwave mobile

Finding hidden frequencies. The rules and guidelines can also be used to locate unknown frequencies. Planning to visit your local airport? The spectrum chart indicates that the civilian air frequencies are between 108.00 and 136.00 MHz. To find the active airport frequencies, you would search between the two ranges.

The radio spectrum

At this point, it should be clear that every user in the radio spectrum must operate their equipment within the guidelines established by the FCC. It doesn't matter if it's an FM radio station, CB radio, airplane, or taxi cab. All users of the radio spectrum must obey strict guidelines that govern the use of radio communications.

Users of the radio spectrum

Table 4-2 shows that the radio spectrum consists of many diversified services. The Forestry Service includes state and national parks, fish & wildlife management, and recreational activities. The Fire Service includes professional and volunteer fire companies. The Police Radio Service is reserved for law enforcement agencies. Private security guards that patrol shopping malls, factories, and college facilities can be found in the Business Radio Service.

Table 4-2. Services that
utilize the radio spectrum

Agency
Aeronautical
Air Force
Amateur radio
Ambulance
Army
Broadcasting
Business
Department of Agriculture
Department of Commerce
Department of Defense
Department of Energy
Department of Justice
Department of the Interior
Experimental
Fire
Forestry
Industrial
Maritime
Navy
National Aeronautics and Space Administration
Police
Satellite
Scientific
Security
Treasury Department

In a later chapter we will provide the frequencies of the more popular services. At this time, we simply want the reader to recognize that each service has its own group of frequencies. A typical example is the Citizens Band Radio Service (CB). The frequencies assigned to CB radio are shown in Table 4-3. A typical CB radio contains 40 channels. Each channel is a "holding place" for a specific frequency. One channel holds one frequency. A 20-channel CB radio or scanner radio would only be capable of holding 20 frequencies.

Table 4-3.
CB radio frequencies

Channel	Frequency
1	26.965
2	26.975
3	26.985
4	27.005

Table 4-3. Continued.

Channel	Frequency
5	27.015
6	27.025
7	27.035
8	27.055
9	27.065
10	27.075
11	27.085
12	27.105
13	27.115
14	27.125
15	27.135
16	27.155
17	27.165
18	27.175
19	27.185
20	27.205
21	27.215
22	27.225
23	27.255
24	27.235
25	27.245
26	27.265
27	27.275
28	27.285
29	27.295
30	27.305
31	27.315
32	27.325
33	27.335
34	27.345
35	27.355
36	27.365
37	27.375
38	27.385
39	27.395
40	27.405

Frequency spacing

Rotate the dial on a standard FM broadcast radio and you'll notice that each number after the decimal is an odd number (Table 4-4). Each frequency is separated by 0.2 MHz. This is called *frequency spacing* and it prevents a radio station from interfering with another nearby station. The frequency separations in the CB radio band prevent a conversation on channel 5 from interfering with someone using channel 6. Additional frequency separations for specific bands are listed in Table 4-5.

Understanding how each band is separated can help you to locate active frequencies. If your local police are using 45.54 MHz, the next available frequency would be 45.56 MHz. Of course, there's no guarantee that the next higher or lower frequency will belong to your local authorities. Small, rural towns may only be authorized to operate on a single frequency. Large cities, such as New York, can be assigned to operate on more than twenty separate frequencies.

Table 4-4. FM radio broadcast band separations

Frequency separations × 0.2
91.10
91.30
91.50
91.70
91.90
92.10
92.30
93.50
93.70
93.90
94.10

Table 4-5. Frequency separations

Frequency	Separation
25.00 to 50.00	.005
108.00 to 136.00	.025
148.00 to 174.00	.005
406.00 to 512.00	.012
806.00 to 912.00	.012

Locating police frequencies

The frequency assignments for your local police are probably listed in scanning directories that are available commercially. *Police Call* (Fig. 4-1) is available from Radio Shack and contains an alphabetical list of cities and frequencies. However, new or sensitive frequencies are not always listed. Hobbyists who understand the rules and guidelines of the radio spectrum can use their knowledge to search for additional, unpublished frequencies.

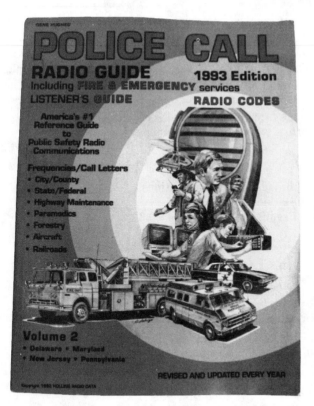

4-1
Police Call is a popular
frequency directory.

AM or FM

The transmitted signal of a CB radio is amplitude modulation (AM). The transmitted signal of an FM radio station is frequency modulation (FM). Explaining the differences between the two carriers requires a lengthy technical explanation that is not necessary. The average hobbyist only needs to know that FM is utilized throughout most of the radio spectrum. The only two services that utilize AM are aeronautical and CB radio.

The AM mode has survived in the aeronautical band for two distinct reasons: (1) AM is the dominant mode used worldwide. Changing to FM on a global scale would require a monumental effort that would cost billions of dollars, and (2) AM circuitry is easier to maintain. The simple circuit design of an AM radio is easy to maintain and repair. Airline companies can't afford to install complex, two-way radios that cannot be quickly repaired or replaced.

False frequencies

All scanner radios are capable of producing *false frequencies*, or images, that are duplicates of the legitimate signal. Suppose that the legitimate aero frequency is 134.400 MHz. Your scanner radio can internally duplicate this frequency and place it on 155.800 MHz. Images are always twice the *intermediate frequency* (IF) of your scanner (Fig. 4-2). Your scanner radio's IF is listed on the specification page of the instruction booklet (Fig. 4-3). All scanners operate on one of the following intermediate frequencies: 10.7, 10.8, or 10.85 MHz.

Scanner radio IF	Multiply ×2	Subtract from image frequency	Actual frequency
10.70	× 2 =	21.40 − 155.80 =	134.40

4-2 To find an image, double the IF frequency of your scanner radio.

SPURIOUS REJECTION:	30 – 54 MHz	50 dB at ⸱ MHz
	108 – 136 MHz	50 dB at 1 MHz
	138 – 174 MHz	50 dB at 1 MHz
	380 – 512 MHz	Not specif
SELECTIVITY:	±9 kHz, −6 dB ±15 kHz, −50 dB	
IF REJECTION: →	10.7 MHz	50 dB at 154 N
SCANNING RATE:	Fast Slow	10 channels/se 5 channels/sec.
SEARCH RATE:	Fast Slow	10 steps/sec. 5 steps/sec.
PRIORITY SAMPLING:	2 seconds	
DELAY TIME:	2 seconds	
MODULATION ACCEPTANCE:	±7 kHz	
IF FREQUENCIES	10.7 MHz and 455 kHz	
FILTERS:	1 crystal filter, 1 ceramic filte	
SQUELCH SENSITIVITY:	Threshold Tight	Less than 1.0 (S+N)/N 15 ⸱
ANTENNA IMPEDANCE:	50 ohms	
AUDIO POWER:	220mW maximum	
BUILT-IN SPEAKER:	2" (5cm) 16 ohm, dynamic t⸱	
POWER REQUIREMENTS:	+9V DC, 6 AA batteries, or a suitable adapter (negative gr⸱ only) Current drain: 55 mA (Squelched) 100 mA (full volume unsou⸱	

4-3 Your scanner radio's instruction booklet contains valuable information.

Images

Images can be discovered by studying the band allocations. If you are hearing aircraft communications on the police bands, don't be misled into believing that you have discovered a new frequency. Simply double your scanner radio's IF and subtract it from the frequency in question. You'll probably discover that the new frequency is actually an image.

Birdies

Birdies are signals that are internally produced by all scanner radios. When your scanner is tuned to a birdie frequency, it will produce a signal without sound. Scanner radio birdie frequencies are usually listed in the instruction booklet. Birdies can also be located by disconnecting your scanner radio from the antenna and searching through each frequency range. Again, when your scanner radio stops on a frequency without sound, that's a birdie. Birdie frequencies are bothersome because they *lock-up* your receiver and prevent it from scanning. The birdie frequencies of your scanner radio should be noted and avoided whenever possible.

Simplex or duplex

When you talk on a cordless phone (Fig. 4-4) you are using a *full duplex* radio system, both parties can speak at the same time. In a simplex system, the first party speaks and the 2nd party listens. The 2nd party cannot respond until the person speaking releases the microphone button. A *simplex* system can utilize one or two frequencies. In a two frequency system, the mobile units talk on one frequency and the base uses the remaining frequency.

4-4
The cordless phone is a full-duplex two-way radio.

American Telegraph and Telephone (AT&T)

Frequency pairs

The majority of radio services that operate between 30.0 and 174.0 MHz utilize a single frequency. The UHF band, 450.0 to 470.0 MHz, utilizes pairs of frequencies that are spaced 5.0000 MHz apart. The UHF-T band, 470.0 to 512.00 MHz also utilizes two frequencies, spaced 3.000 MHz apart.

In the UHF and UHF-T bands, the lower frequency is assigned to the base station. The higher frequency is used by mobile units. Frequencies between 806.00 and 941.00 MHz are always paired and separated by 45.0000 MHz. The higher frequency is the base and the lower is for mobile use.

Types of stations

A *base station* (Fig. 4-5) transmits directly to mobile and portable units. Base stations that are located away from the dispatcher (Fig. 4-6) are known as *remote base stations*. Remote base stations are often controlled via telephone lines or by microwave signals. If the mobile units are at a considerable distance from the base unit, a repeater (Fig. 4-7) may be utilized.

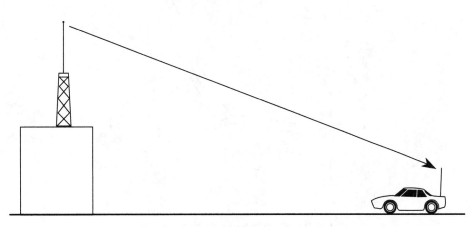

4-5. Example of a base station.

Repeaters

Repeaters are used to extend the range of mobile and portable communications. Repeaters are also used to eliminate *dead spots* and to provide full radio coverage within a specific area (Fig. 4-8). Repeaters receive on one frequency (input) and retransmit the signal on another frequency (output). Scanner buffs can hear both sides of the conversation by monitoring the repeater output frequencies.

Mobile repeaters

Mobile extenders or *vehicular repeaters* (Fig. 4-9) utilize the vehicle's mobile radio to increase the range of handheld radios. Mobile extenders receive the low

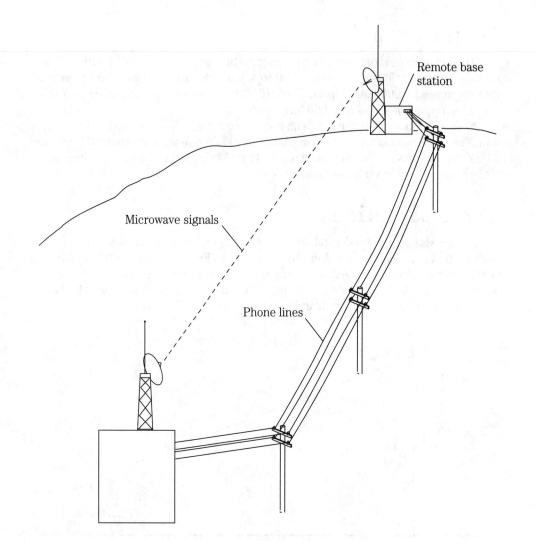

4-6 Operation of a remote base station.

power signal from portable units and use the high power vehicular radio to re- transmit the signal. Mobile extenders provide reliable communications to law enforcement officers who are required to exit their vehicles must use portable radios.

Narrow or wide band

Commercial users of the radio spectrum are usually assigned frequencies in the FM wide band. Narrow Band FM (NBFM or NFM) is used for approximately 90% of all two-way radio communications. Table 4-6 shows the narrow and wide band frequency assignments.

Table 4-6. Narrow and wide band frequency assignments

Narrow band	Wide band
29.00 to 54.00	54.00 to 72.00
72.00 to 76.00	76.00 to 108.0
136.00 to 174.00	175.00 to 216.00
217.00 to 225.00	226.00 to 400.00
406.00 to 512.00	401.00 to 405.00
	512.00 to 806.00

Note: Most scanner radios will select the proper mode automatically. There are a few models, however, that require the user to manually set the correct band width.

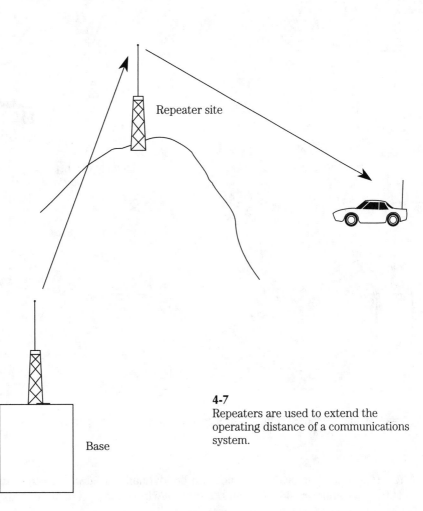

Repeater site

Base

4-7
Repeaters are used to extend the operating distance of a communications system.

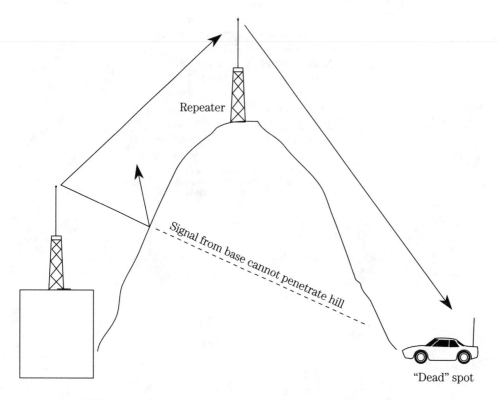

Repeater

Signal from base cannot penetrate hill

"Dead" spot

4-8 Strategically placed repeaters can help to eliminate dead spots.

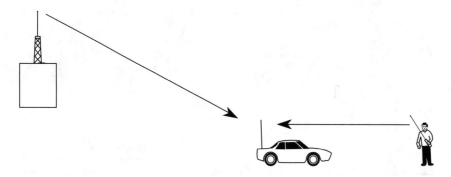

4-9 Mobile extenders increase the signal strength of handheld transmitters.

A few top-of-the-line scanner radios can be manually programmed to scan the FM wide band. Scanner radios without the narrow/wide band option are designed at the factory to automatically receive narrow band FM.

Voice security

As the popularity of scanning continues to grow, security-conscious agencies are beginning to use a variety of voice protection techniques. The most common and inexpensive method is the 10 code (Table 4-7). To further discourage monitoring, some agencies will use personalized codes that are not available to the public.

Table 4-7. Standard radio codes

Commonly used codes

Ten-1	Repeat your message.
Ten-2	Signal is loud and clear.
Ten-3	Stop transmitting.
Ten-4	Okay.
Ten-5	Relay information.
Ten-6	Busy.
Ten-7	Out of service.
Ten-8	In service.
Ten-9	Repeat message
Ten-10	Negative, no.
Ten-11	_____in service.
Ten-12	Stand-by.
Ten-13	Report conditions.
Ten-14	Standby for information.
Ten-15	Message delivered.
Ten-16	Reply to message.
Ten-17	En route to scene.
Ten-18	Urgent.
Ten-19	Contact_____.
Ten-20	Unit location.
Ten-21	Call by telephone.
Ten-22	Cancel.
Ten-23	Arrived at scene.
Ten-24	Assignment complete.
Ten-25	Meet with_____.
Ten-26	Estimated time of arrival.
Ten-27	License information.
Ten-28	Vehicle registration information.
Ten-29	Check records.
Ten-30	Use caution.
Ten-31	Pick up.
Ten-32	Additional units requested.
Ten-33	Emergency.
Ten-34	Correct time.

Note: The above codes are officially suggested by the Associated Public Safety Communications Officers (APCO). Many communities change the codes to suit their particular needs.

Voice security can also be achieved by utilizing electronic devices that scramble or encrypt the radio signal. The most commonly used method is known as speech inversion. This scrambling technique inverts the high and low tones of the human voice. A scrambled message using voice inversion sounds garbled, similar to the voice of a cartoon duck. Prior to 1985, speech inversion decoders were available for less than 50 dollars. At this writing, decoding devices for scanner radios are illegal and cannot be sold to the general public. (Scanning laws will be thoroughly discussed in chapter 5.)

Protecting sensitive communications

To protect sensitive communications belonging to the FBI, Secret Service, and other federal agencies, the U. S. government utilizes the "Data Encryption Standard" (DES). DES converts voice communications into a digital code. When encountered on the air, Digital Voice Protection (DVP) sounds like an increase in background noise. No voice patterns are heard.

Developed by Motorola, DVP rearranges the voice components into millions of digital pieces. Chance decoding, even with a computer, is virtually impossible. Digital encryption is available for federal and nonfederal use. It's not uncommon for police departments in large cities to utilize digital encryption to protect surveillance and other discrete communications.

Breaking the rules

Can the rules be broken? The answer is a resounding yes! Federal law enforcement agencies, especially the FBI and Secret Service, will occasionally utilize frequencies that are not assigned for federal use.

Temporary communications during presidential visits are a typical example. During a visit to New York City, the President decided to take an unscheduled train ride. The Secret Service responded by borrowing two-way radio equipment owned by the railroad. The railroad communications system provided the Secret Service with reliable radio service that could be utilized during an emergency.

The Federal Bureau of Investigation (FBI) has also conducted surveillance operations on nonfederal frequencies. Federal agents will usually select a business frequency that is not utilized after closing hours. An ideal candidate would be a radio dispatched delivery service that closes at 5 PM. When the frequency becomes vacant, it is simply borrowed and used for federal communications.

Rental frequencies

Rental radio is a relatively new concept that is growing in popularity. Rental companies provide businesses and promoters of special events with rented two-way equipment.

Several months prior to the event, rental agents will visit the area to determine the frequencies that can be borrowed. Frequencies are usually borrowed from local area businesses. Upon FCC approval, the rental company prepares their rental equipment to operate on the borrowed frequencies.

During special events (Fourth of July celebrations, large parades, air shows, etc.) it may be possible to hear private security guards, television crews, and support staff on frequencies that belong to other agencies. Upon completion of the special event, the rental radios are returned and the borrowed frequencies resume normal operations.

Sit back and listen

Rules, guidelines, frequency separations, AM/FM, images, birdies, repeaters, and more . . . wow! Did you understand everything? If you're not comfortable with any of the topics that were presented in this chapter, don't be too concerned. You can enjoy the hobby of scanning without memorizing a single rule or regulation.

For now, it's perfectly acceptable to simply sit back and listen. As you become an experienced hobbyist, the rules and guidelines will become an invaluable reference source that will help to develop and improve your skills.

5
CHAPTER

Scanning laws

When a scanner radio is used to monitor two-way communications, the user is considered to be the *third party* listener (Fig. 5-1). The first party is the person talking into the microphone and the second party is the intended recipient of the message.

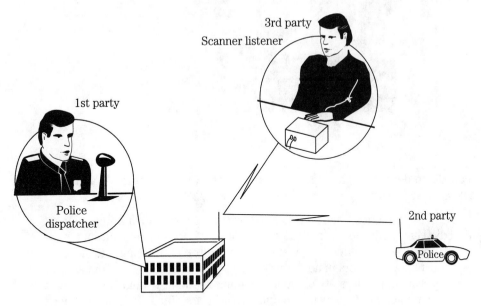

5-1 Scanning hobbyists are third party listeners.

In 1934, the Communications Act established specific laws for controlling 3rd party monitoring. The laws prohibit 3rd party listeners from revealing the content of the intercepted transmission. In addition, 3rd party listeners cannot use the information for personal gain. A vehicle towing service, for example, cannot monitor po-

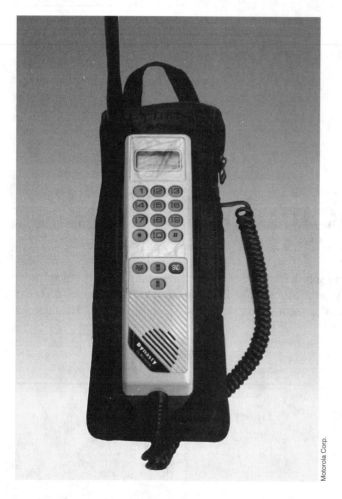

Motorola Corp.

5-2
A cellular car phone is
nothing more than a
two-way radio.

lice calls to determine the location of accidents. A portion of the Communications
Act reads as follows:

> No person not being authorized by the sender shall intercept any ra-
> dio communication and divulge or publish the existence, contents, sub-
> stance, purport, effect, or meaning of such intercepted communication to
> any person. No person not being entitled thereto shall receive or assist in
> receiving any interstate or foreign communication by radio and use such
> communication (or any information therein contained) for personal ben-
> efit or for the benefit of another not entitled thereto. No person having
> received any intercepted radio communication or having become ac-
> quainted with the contents, substance, purport effect, or meaning of such
> communication (or any part thereof) knowing that such communication
> was intercepted, shall divulge or publish the existence, contents, sub-
> stance, purport, effect, or meaning of such communication (or any part
> thereof), or use such communication for his own benefit or for the bene-
> fit of another not entitled thereto.

Understanding the law

In summary, the Communications Act is easily defined. "Don't talk about what you hear!" It's perfectly okay to listen until you are content. It is illegal, however, to discuss your monitoring activities with your neighbor.

The battle between the right to privacy and the right to listen did not surface again until 1985. Technology was growing by leaps and bounds, and there seemed to be an endless supply of new and innovative electronic devices. The popularity of one particular device, the cellular phone (Fig. 5-2), was growing in popularity. Steadily declining prices and the need for reliable roadside communications prompted millions of Americans to purchase cellular phones.

The majority of cellular customers failed to realize that a car phone was nothing more than a two-way radio. Situated between 870.00 and 890.00 MHz were conversations between elected officials, salesmen, lovers, criminals, and thousands more. Anyone with a cellular-capable radio could tune in husbands cheating on their wives, government officials accepting bribes, and drug dealers making international connections. When the Mafia began to use cellular phones, law enforcement agencies became regular cellular listeners.

The Electronic Communications Privacy Act (ECPA)

Public interest in cellular monitoring was fueled by newspapers, magazines, and television news broadcasts that exaggerated and exploited the issue. The cellular industry responded by lobbying for a new monitoring law. The Electronic Communications Privacy Act (ECPA) of 1986 declared the cellular frequencies off limits to third party listeners. For the first time in history, it was illegal to monitor a specific portion of the radio spectrum. The following paragraphs are an excerpt from the ECPA.

2511. Interception and disclosure of wire, oral, or electronic communications are prohibited. Except as otherwise specifically provided in this chapter it is unlawful for any person to:

1a. Intentionally intercept, endeavor to intercept, or procure any other person to intercept or endeavor to intercept, any wire, oral, or electronic communications.

b. Intentionally use, endeavor to use, or procure any other person to use or endeavor to use any electronic, mechanical or other device to intercept any oral communication when

b.(1) Such a device is affixed to, or otherwise transmits a signal through, a wire, cable, or other like connection used in wire communications or

b.(2) Such device transmits communications by radio, or interferes with the transmission of such communication or

b.(3) Such person knows, or has reason to know, that such device or any component thereof has been sent through the mail or transported in interstate or foreign commerce.

The implications of the ECPA are not easily summarized. The law states that anyone caught listening to the cellular car phone frequencies can be arrested! It is also a violation of the ECPA to unscramble encrypted transmissions and to monitor voice paging services. Admittedly, the law is difficult, if not impossible, to enforce. Unless the listener confesses to specific violations, there's no way to prove that someone was listening.

Cellular phone monitoring

Ironically, the ECPA actually helped to increase the interest in cellular monitoring. It was similar to posting a wet paint sign. It invited the curious. People who had never monitored a two-way radio communication were purchasing scanner radios and listening to the cellular frequencies. Budget conscious individuals, who didn't want to spend several hundred dollars on a scanner radio, monitored the cellular frequencies on their television sets.

Table 5-1 shows that the cellular and television frequencies are separated by a very narrow margin. The wide tuning range of a television will allow strong, local cellular calls to be heard on the TV set's audio. To test the idea, simply connect a UHF antenna to your TV and use the remote to tune channels 70 through 83 (Fig. 5-3).

**Table 5-1. Cellular phone
and television frequencies**

Cellular phone frequency	Television channel	Frequency
870.00 to 890.00	#80	866.00 to 872.00
890.00 to 896.00	81	872.00 to 878.00
	82	878.00 to 884.00
	83	884.00 to 890.00

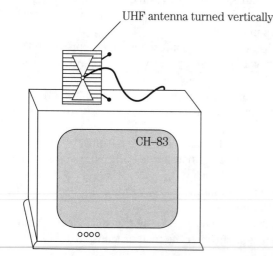

UHF antenna turned vertically

CH–83

5-3
You can listen to cellular conversations with your television set.

Reaction to the ECPA

A mixed reaction to the ECPA by equipment manufacturers added to the confusion. A few companies voluntarily deleted the cellular frequencies from their scanner radios. Others chose to capitalize on consumer interest by providing after-market frequency converters. These small, battery operated devices converted the 450- to 512-MHz UHF band to receive 800-MHz communications. Because most scanning receivers included the UHF band, anyone with an inexpensive converter could monitor the cellular frequencies.

Additional scanning laws

The lack of compliance and the inability to enforce the ECPA set the stage for additional anti-scanning laws. The need for a tougher law was prompted by the discovery of cellular-restorable scanner radios. In specific models, the cellular frequencies could be restored by removing the radio's case and cutting a single wire or diode. A comprehensive list of manufacturers and model numbers that was widely circulated helped the hobbyist to target specific receivers. Additional interest in the subject was fueled by several books that explained and illustrated hundreds of radio modifications. With nothing more than a few basic hand tools, scanning hobbyists learned how to customize a scanner radio (Fig. 5-4).

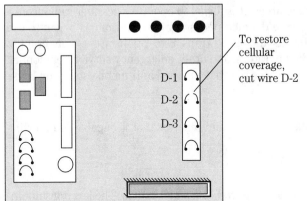

To restore cellular coverage, cut wire D-2

5-4
Cutting a single wire can restore cellular capability to some scanner radios.

Cellular modifications

In an attempt to make cellular modifications more difficult, the cellular industry successfully lobbied for passage of the Telephone Disclosure and Dispute Resolution Act, which was passed in April of 1993. It prohibits the manufacture or importation of cellular-capable or cellular-restorable scanner radios and converters after April of 1994. The law requires manufacturers to produce receivers that cannot be easily modified to receive cellular phone calls. Scanning receivers that include the cellular frequencies from the factory cannot be sold in the United

States. Low cost converters, which permit cellular monitoring on a different band, have also been deemed illegal.

Canadian cellular laws

The concern for cellular privacy has also spread into Canada. Amendments to Canada's criminal code and Radio Communications Act were enacted after a number of embarrassing cellular phone calls were made public. A British Columbia cabinet minister resigned and a Quebec Premier's credibility was destroyed after their cellular phone calls became public knowledge. To make matters worse, England's Princess Diana was also monitored on a cellular phone. Her conversations to a male friend were widely published in the United States and abroad.

Under Canadian law, it is illegal to divulge the contents of a cellular phone call. It is also illegal for third parties, including radio and television stations, to repeat or print the contents of a cellular call. In an effort to educate the public, Canada has issued a small pamphlet containing privacy information. Cellular users are advised that cellular phones are wireless devices that can be monitored. Scanner enthusiasts are advised that violations of Canada's Radio Communication Act will be prosecuted.

State laws

In addition to the new federal laws, the hobby of scanning is also affected by local city and state laws. In certain areas, it is illegal to operate a scanner radio from a vehicle. Mobile scanning laws were deemed necessary when criminals began to use scanner radios to avoid road blocks and speed traps. Exceptions are usually granted to volunteer firemen, ham radio operators, and other emergency services personnel. The following list represents the states that have passed anti-mobile scanning laws:

Florida

The law prohibits the installation of a radio receiver that is capable of receiving police communications.

Indiana

The law prohibits the possession of portable scanner radios. The law does not include base sets that are designed for use in a building or dwelling.

Kentucky

Has passed the most restrictive monitoring law in the United States. The possession of a radio that is capable of receiving police communications is illegal. Transporting a scanner radio in your trunk is considered to be a violation. Ham radio operators are not exempt.

Michigan

It is illegal to equip a vehicle with a scanner radio.

Minnesota

The use of a scanner in a motor vehicle is illegal.

New York

Installing a mobile radio that can receive police communications is illegal. Portable scanner radios are not specifically mentioned.

Rhode Island

Convicted felons are prohibited for life from possessing any type of scanner radio.

South Dakota

Convicted felons within the past ten years are prohibited from possessing any type of scanner radio.

Readers should note that new laws are proposed and old laws are amended on a daily basis. Do not install or transport a scanner radio based solely upon what has been presented in this chapter. It is your responsibility to check local and state laws.

Cordless phone laws

Cordless phone monitoring is another aspect of the hobby that has provoked the passage of additional anti-scanning laws. Cordless phones, like cellular phones, utilize an antenna to transmit a two-way conversation (Fig. 5-5). Although the cordless signal is very weak, it can often be monitored from several blocks to several miles away. To discourage cordless monitoring, the state of California has passed a law that makes cordless monitoring illegal. The actual frequencies that should be avoided are provided in chapter 6.

The law and you

As pointed out, the hobby of scanning is regulated by the following: Communications Act of 1934, the ECPA of 1986, the Telephone Disclosure and Dispute Resolution Act of April 1993. Depending on your location, state and/or city laws may also apply.

At first glance, the various laws may seem overwhelming. To clarify the situation, the hobbyist should focus on one simple guideline: don't repeat what you have monitored. The same guideline can also be applied to mobile listening, but don't forget that mobile scanning is often regulated by state and local laws.

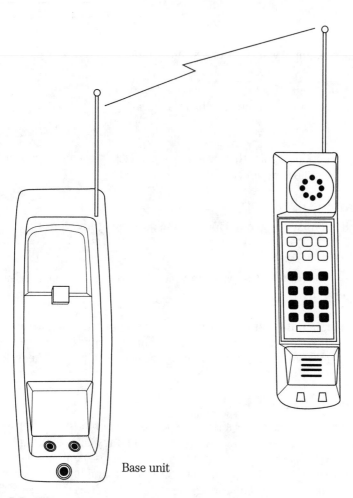

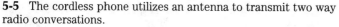
Base unit

5-5 The cordless phone utilizes an antenna to transmit two way radio conversations.

The need for more laws

Although it is impossible to predict the future, we doubt that additional anti-scanning laws will be necessary. Within a few years, the cellular industry is expected to begin converting to a digital system. The new technology will separate the cellular signal into millions of digital codes. A microcomputer in the cellular phone will instantly decode the signal and provide clear voice to the cellular customer. Because third party monitoring will be impossible, the need for additional anti-scanning laws will rapidly diminish.

Enforcement

As mentioned earlier, the anti-scanning laws are very difficult to enforce. It would be remiss however, to fail to report that several listeners have been prosecuted. In each case, the third party listener violated the ECPA by tape recording cellular phone calls. The tapes were sent to television and newspaper reporters and the actual conversations were made public. When multi-million dollar lawsuits were filed, confidential informants provided the police with the name and address of the person(s) that made the actual tape.

Remember one rule

There's no need to worry about breaking the rules. You won't go to jail and you won't get into trouble if you observe one simple rule. And redundant as it might seem, it's necessary to repeat it one last time: listen to your heart's content, but don't repeat what you hear.

6

CHAPTER

Popular monitoring targets

The public safety frequencies are the number one monitoring target in the nation. Everyone wants to monitor their local police, fire, and ambulance frequencies. When emergency sirens pierce the air, your scanner radio will provide a front row seat to all the action. You'll hear high-speed police pursuits, criminals being apprehended, and fire fighters rushing to fire scenes.

The exact frequencies for your area can be located by utilizing a variety of techniques. The preferred method is to use a frequency directory. The number one choice among scanner buffs is Radio Shack's *Police Call*. The directory is arranged in alphabetical order and provides local frequencies for thousands of communities. Joining a scanning club is another excellent way to obtain local frequencies. Scanning clubs are located in nearly every state and they are operated by fellow hobbyists. A list of scanning clubs is provided in chapter 10.

Exploring your world

Active frequencies can also be discovered from the privacy of your home. Nothing can compare to sitting in front of your scanner radio and listening for an unexpected adventure. You become an explorer in an invisible world filled with frequencies, mystery, and intrigue. To make your adventure more enjoyable, we have listed and briefly described a few of the popular agencies that you can monitor.

Amateur radio

Amateur radio operators (hams) can be monitored on a variety of different frequency bands. The popular bands that scanner hobbyists can monitor are shown in Table 6-1. These frequencies are used primarily for local communications and are used by hams to discuss a variety of topics. Larger ham clubs will conduct monthly meetings that you can attend. The date and time of the meetings will be announced on the air. Don't be bashful, your attendance at a local ham meeting could inspire you to become a licensed ham.

Table 6-1.
Amateur radio frequencies

Ham frequencies	Wavelength
28.00 to 29.70	10 meters
50.00 to 54.00	6 meters
144.00 to 148.00	2 meters
222.00 to 225.00	1.25 meters
420.00 to 450.00	70 centimeters
902.00 to 928.00	30 centimeters
1240 to 1300.00	23 centimeters

Note: Amateur radio operators refer to individual bands according to wavelength. The term *meters* refers to a particular band of frequencies.

Baby monitors

Parents with small children will often monitor the youngster's bedroom with a baby monitor. The transmitter is powered by house current and it is typically placed in the child's room. The receiver is battery powered and can be carried to any convenient listening area.

Baby monitors transmit a low-power signal that can be received several blocks away. Table 6-2 shows the five frequencies that can be monitored. The transmitter is actually a self-installed bugging device that is rarely turned off. As the transmitter is moved about the house, you may hear conversations and sounds that are normally associated with X-rated movies!

Table 6-2. Baby
monitor frequencies

Frequencies
49.83
49.845
49.86
49.875
49.89

Business communications

Radio communications provide a vital service to the business community. From large corporations with thousands of employees, to small family-run establishments, the business radio frequencies are alive with activity. The active business frequencies in your area can be discovered by searching through the business frequency ranges in Table 6-3.

Table 6-3.
Business frequencies

Frequency ranges to search
33.00 to 46.00
150.80 to 162.00
461.00 to 465.00
502.00 to 512.00
851.00 to 853.00

It is impossible to list all the various business communications that will be heard. Plumbers, carpenters, electricians, security guards, limousine services, taxi cabs, parcel delivery services, trucking companies, and thousands more can all be monitored on the business frequencies.

Itinerant frequencies

Hidden within the business bands are a small group of *itinerant frequencies* (Table 6-4). The frequencies are limited to low-power applications in factories, small businesses, and special events. The itinerant frequencies are often used in parades, balloon races, traveling shows, and other short-lived events. The itinerant frequencies are full of surprises. A single frequency can be used to coordinate a balloon race in the morning, to provide parade security in the afternoon, and to launch fireworks in the evening.

Table 6-4. Itinerant frequencies

Frequencies
27.49, 35.04, 43.04, 151.505, 151.625, 158.40, 451.80, 456.80, 464.50, 464.55

Colored dot frequencies

Colored dots are used to designate a few specific frequencies in the Business bands. The dot frequencies are shown in Table 6-5. To monitor a blue dot frequency, you would enter 154.57 MHz into your scanner radio. Colored dots are often affixed to handheld business radios. The colored dots make it easy to assign a specific frequency to a large group of users. Team A could be assigned blue dot radios, and team B could be assigned green dot radios. It's quick and fool-proof, and the user doesn't need a technical background to coordinate several different frequencies.

Table 6-5.
Colored dot frequencies

Frequencies	
151.625	Red dot
154.57	Blue dot
154.60	Green dot

Table 6-5. Continued.

Frequencies

464.50	Brown dot
464.55	Yellow dot
167.73	Black dot

News media

Millions of Americans watch the news on TV, or listen to it on their favorite radio station. You can hear the news live on your scanner radio. The news frequencies used by reporters are in Table 6-6. News vans, helicopters, and traffic reporters are just a few of the communications that can be monitored. It's not uncommon to hear reporters rehearsing their lines, or to hear technicians checking the quality of the signal. To find additional frequencies that are not listed, search between the following: 450.05 to 450.925 and 455.05 to 455.925 MHz.

Table 6-6. News frequencies

Paired frequencies are as follows:

Output	Input	Output	Input
450.05	455.05	450.40	455.40
450.0875	455.0875	450.4125	455.4125
450.10	455.10	450.4375	455.4375
450.1125	455.1125	450.45	455.45
450.1375	455.1375	450.4625	455.4625
450.15	455.15	450.4875	455.4875
450.1625	455.1625	450.50	455.50
450.1875	455.1875	450.5125	455.5125
450.20	455.20	450.5375	455.5375
450.2125	455.2125	450.55	455.55
450.2375	455.2375	450.5625	455.5625
450.25	455.25	450.5875	455.5875
450.2625	455.2625	450.60	455.60
450.2875	455.2875	450.70	455.70
450.30	455.30	450.75	455.75
450.3125	455.3125	450.80	455.80
450.3375	454.3375	450.85	455.85
450.35	455.35	450.90	455.90
450.3625	455.3625	450.925	455.925
450.3875	455.3875		

Rail traffic

The national railroad frequencies are shown in Table 6-7. Railroad communications are vital to the economy and can be monitored around the clock. It's possible to hear track maintenance crews, switching towers, inspectors, and railroad police officers. Scanning hobbyists in large cities should also search for local commuter and freight train frequencies.

Table 6-7. Railroad frequencies

Train frequencies

159.81, 159.93, 160.05, 160.185, 160.20, 160.215, 160.23, 160.245, 160.26, 160.275, 160.29, 160.305, 160.32, 160.335, 160.35, 160.365, 160.38, 160.395, 160.41, 160.425, 160.44, 160.455, 160.47, 160.485, 160.50, 160.515, 160.53, 160.545, 160.56, 160.575, 160.59, 160.605, 160.62, 160.635, 160.65, 160.665, 160.68, 160.695, 160.71, 160.725, 160.74, 160.755, 160.77, 160.785, 160.80, 160.815, 160.83, 160.845, 160.86, 160.875, 160.89, 160.905, 160.92, 160.935, 160.95, 160.965, 160.98, 160.995, 161.01, 161.025, 161.04, 161.055, 161.07, 161.085, 161.10, 161.115, 161.130, 161.145, 161.160, 161.175, 161.19, 161.205, 161.22, 161.235, 161.250, 161.265, 161.280, 161.295, 161.31, 161.325, 161.34, 161.355, 161.37, 161.385, 161.40, 161.415, 161.430, 161.445, 161.460, 161.475, 161.490, 161.505, 161.52, 161.535, 161.55, 161.565

Fast food

"Two hamburgers and an order of fries." Scanning your local McDonald's order window may not be very exciting, but it can be a novel experience. Table 6-8 contains the frequencies for several fast food chains. The signal power is very low and rarely travels more than a few hundred feet. If your monitoring efforts are to be successful, it might be necessary to take your scanner radio to the fast-food parking area.

**Table 6-8.
Fast food frequencies**

Frequencies
30.84
31.01
33.14
35.02—McDonald's base
151.895
154.57
154.60
170.245
170.305
171.105
457.550
457.60

Table 6-8. Continued.

Frequencies
460.8875
467.775
467.825

Air adventures

Civilian airports are also fun to monitor. Each individual airport service (Table 6-9) operates on a separate frequency. Ground control, approach control, and departure control are just a few of the services that are constantly in use. Idle chit-chat between pilots, although frowned upon by the FCC, can be monitored on 122.850, 122.900, 122.925, and 123.100 MHz.

Table 6-9. Civilian airport frequencies

Frequencies	
118.00 to 121.40	Airport towers
121.50	Emergency
121.6 to 121.95	Ground control
121.975 to 122.675	Flight service
122.70 to 122.825	UNICOM—private airport information
122.975 to 123.075	UNICOM—Helicopter
123.60 to 123.650	Arrival and departure
123.675 to 128.80	Air Traffic Control
128.225 to 132.00	Airline company communications
132.025 to 135.975	Air Traffic Control

Aircraft repair frequencies

Another group of aircraft frequencies are shown in Table 6-10. They are commonly referred to as the *write-up* frequencies. They are used by pilots to report equipment failures and to request aircraft maintenance. You'll probably be surprised by the number and variety of daily repair requests.

**Table 6-10.
Aircraft repair frequencies**

Repair frequencies
129.30
129.70
130.25
130.60
130.65
130.85

Table 6-10.Continued.

Repair frequencies
131.150
131.425
132.00

Military air

Military airports can also provide hours of fascinating listening. You'll hear mock air battles, in-flight refueling operations, training exercises, and much more. The military air frequencies are shown in Table 6-11. It's important to remember that both military and civilian air frequencies are AM communications. If your scanner radio doesn't automatically switch to the AM mode, don't forget to manually select the proper mode.

Table 6-11.
Military air frequencies

Military air frequencies	
236.60	Military control towers
237.90	Coast Guard aircraft
243.00	Emergency
255.40	Flight service
277.80	Naval fleet aircraft
282.80	Search and Rescue
311.00	Strategic Air Command
340.20	Naval control towers
372.20	Pilot to tower
381.80	Coast Guard aircraft

Aero phones

Our expanding technology has made it possible to make phone calls from airplanes. Phone calls from privately owned aircraft can be monitored between 459.70 and 459.975 MHz. Phone calls from commercial airplanes can be monitored between 894.00 and 896.00 (AM mode) MHz.

The airline phone call is transmitted to a ground station that connects the radio signal to a land line phone system. You'll probably only hear the air-to-ground side of the conversation. Ground stations, which transmit the ground-to-air side of the conversation, are widely scattered and can't be monitored from long distances.

Cordless microphones

Cordless microphones can be found in nearly every city across America. From Broadway productions to one night stands, cordless microphones are the preferred choice among amateur and professional performers. Cordless microphones are nothing more than small, handheld transmitters that you can monitor. The low-power signal won't travel very far, but if you live near a night club or bar it might be possible to monitor a live performance. The frequencies for cordless microphones are shown in Table 6-12.

Table 6-12. Cordless microphone frequencies

Frequencies
169.505
170.245
170.305
171.045
171.105
171.845
171.905

Can't get tickets to your favorite concert? Take your scanner radio to the parking lot and scan the cordless microphone frequencies. If the performers are using cordless microphones, the entire performance can be heard on your scanner. The sound quality won't be that great, but it's better than missing the entire show!

Cordless phones

The cordless phone is another popular gadget that is nothing more than a two-way radio. The cordless base unit transmits a radio signal that can be monitored up to several miles away. Cordless owners usually fool themselves into believing that the monitoring distance is directly related to the working distance of the phone. Few people realize that a scanner radio can detect and amplify a cordless signal from a mile or more away. The cordless phone frequencies are shown in Table 6-13. To hear both sides of the conversation, you'll need to monitor the cordless base frequencies.

Table 6-13. Cordless phone frequencies

	Frequencies
Channel 1	46.61 base/49.67 handset
2	46.63 base/49.85 handset
3	46.67 base/49.86 handset
4	46.71 base/49.77 handset
5	46.73 base/49.875 handset
6	46.77 base/49.83 handset
7	46.83 base/49.89 handset
8	46.87 base/49.93 handset
9	46.93 base/49.99 handset
10	46.97 base/49.97 handset

The popularity of cordless monitoring has prompted the state of California to pass an anti-cordless monitoring law. The law states that it is illegal to monitor and/or print the contents of a cordless conversation. Additional state and federal laws are presented in chapter 5.

Fire departments

A small sampling of fire department frequencies are listed in Table 6-14. It is important to remember that the fire dispatch frequencies are always separate from the frequencies that are used on the actual fire scene. The fire scene frequencies are assigned to low-power, handheld radios that firemen carry into buildings. If you're more than two miles away from the fire, it may be impossible to monitor the fire scene action.

**Table 6-14. Partial listing
of fire department frequencies**

Fire Dept. frequencies
46.06
46.08
46.10
46.12
46.14
46.16
46.18
46.20
46.22
46.24
46.26
46.28
46.30
46.32
46.34
46.38
46.40
46.42
46.44
46.46
46.48
46.50
Also search: 154.01 to 154.445

Medivac frequencies

Medivac frequencies are used to relay vital patient information between paramedics and hospitals. It is not uncommon for two separate frequencies to be used. To hear both sides of the conversation, it might be necessary to monitor the paired frequencies listed in Table 6-15. Other frequencies that may be active are 155.16, 155.28, and 155.34 MHz.

Table 6-15.
Paramedic frequencies

Medical frequencies		
462.950	paired with	462.9625
462.975	"	462.9875
463.00	"	463.0125
463.025	"	463.0375
463.050	"	463.0625
463.075	"	463.0875
463.100	"	463.1125
463.125	"	463.1375
463.150	"	463.1625
463.175	"	463.1875

Monitoring the President

Presidential security is the responsibility of the Secret Service. When the President, Vice President, or family members depart the White House, Secret Service Agents are never far away. The Secret Service frequencies are provided in Table 6-16. Individual code names, used by the Secret Service to identify members of the Presidential family, are shown in Table 6-17.

Table 6-16.
Secret Service/Presidential security frequencies

Secret Service frequencies	
164.8875	
165.213	
165.65	
165.785	
166.213	
166.40	
166.70	
167.7875	
169.925	
407.825	Concealed radio transmitters
407.875	Concealed radio transmitters

Table 6-17. Secret Service code names

Field code names	Description
Buckeye	Camp David
Caravan	Vice President's support vehicle
Cargo	First Lady's vehicle
Challenger	Presidential night security
Crown	White House
Curbside	Washington, DC, airport
Falcon	Tactical response team
Halfback	President's support vehicle
Hercules	Sniper response team
Stage Coach	Presidential limo

If the President visits your local area, the Secret Service frequencies will become active several days prior to the actual arrival date. Secret Service communications will contain a substantial amount of Digital Voice Protection (DVP). We discussed DVP and other voice scrambling techniques in chapter 5. When DVP is used, it sounds like an increase in background noise, and voice characteristics cannot be recognized.

Monitoring federal agents

Federal law enforcement agents can also be monitored. The Federal Bureau of Investigation (FBI) has field offices in nearly every large city in America. Table 6-18 shows a few of the nationwide frequencies that have been assigned to the FBI. Don't expect them to utilize each frequency that is provided. You'll need to monitor the entire list and select the frequencies that are active in your area.

Table 6-18. FBI frequencies

FBI	
162.6375	164.40
162.7625	164.80
162.95	165.25
162.975	166.50
163.10	167.15
163.30	167.2125
163.625	167.225
163.65	167.425
163.985	167.4375
164.275	167.50

Military installations

Military base operations are also assigned to a specific group of frequencies. Hobbyists who live within 50 miles of a military installation can monitor the frequencies contained in Table 6-19. Large military bases are self-supporting communities that have internal police, fire, and ambulance services. Hobbyists can also monitor building maintenance crews, vehicle repair services, and hundreds of additional communications that are vital to the mission of the base.

Table 6-19. Military base frequencies

148.00	149.00
148.05	149.05
148.10	149.10
148.15	149.15
148.20	149.20
148.25	149.25
148.30	149.30
148.35	149.35
148.40	149.40
148.45	149.45
148.50	149.50
148.55	149.55
148.60	149.60
148.65	149.65
148.70	149.70
148.75	149.75
148.80	149.80
148.85	149.85
148.90	149.90
148.95	149.95

Ocean adventures

The frequencies for commercial ships and pleasure boats are listed in Table 6-20. During the summer months, these frequencies are alive with recreational boating activities. Commercial ships, tug boats, and the Coast Guard can be monitored on a yearly basis. The frequencies are not limited to coastal areas. Hobbyists living near large rivers, lakes, and other inland waterways will discover that the marine frequencies are used in these areas as well.

Table 6-20. Maritime frequencies

Frequency	Channel	Purpose
156.30	6	Safety
156.35	7	Commercial
156.40	8	Commercial
156.45	9	Commercial
156.50	10	Commercial
156.55	11	Commercial
156.60	12	Port operations
156.65	13	Navigational
156.70	14	Port operations
156.75	15	Environmental
156.80	16	Distress
156.85	17	State control
156.90	18	Commercial
156.95	19	Commercial
161.60	20	Port operations
157.05	21	Coast Guard
157.10	22	Coast Guard
157.15	23	Coast Guard
161.80	24	Marine phone calls
161.85	25	Marine phone calls
161.90	26	Marine phone calls
161.95	27	Marine phone calls
162.00	28	Marine phone calls
156.275	65	Port operations
156.325	66	Port operations
156.375	67	Commercial
156.425	68	Pleasure boating
156.475	69	Pleasure boating
156.525	70	Pleasure boating
156.575	71	Pleasure boating
156.625	72	Pleasure boating
156.675	73	Port operations
156.725	74	Port operations
156.875	77	Commercial
156.925	78	Pleasure boating
156.975	79	Commercial
157.025	80	Commercial
157.175	83	Coast Guard Auxiliary
161.825	84	Marine phone calls
161.875	85	Marine phone calls
161.925	86	Marine phone calls
161.975	87	Marine phone calls
157.425	88	Commercial

Outer space

Your scanning abilities are not limited to earthly adventures. The space shuttle frequencies in Table 6-21 are waiting to take you into outer space. Your reception of shuttle communications is limited by the orbiting location of the shuttle and by your relative position on Earth. The ideal time to monitor the shuttle is when it is passing directly above your location. Of course, everyone doesn't live within the orbiting path and there's no guarantee that a radio transmission will occur during the optimum time.

Table 6-21. Space shuttle frequencies

Frequencies
259.70 (AM mode)
296.80 (AM mode)
Russian orbiting spacecraft can also be monitored on:
143.625, 166.135

The complications that accompany space shuttle monitoring are considered to be part of the challenge. During shuttle missions, hobbyists throughout the nation will attempt to beat the odds and monitor an outer space adventure. To increase your chances for success, we've also included the frequencies for Russian spacecraft missions. Good luck!

Satellites

Satellite communications are another fascinating aspect of our hobby. Table 6-22 shows the satellite frequencies that can be monitored. Each hour, especially during the evening, the ATS-3 (Applied Technology Satellite) transmits a signal to Earth. The ATS-3 links schools, medical facilities, and scientific expeditions.

Table 6-22. Satellite frequencies

Frequencies
Search between 135.575 and 135.625 for clear voice. Radio beacons can be monitored on the following:

136.77	137.62
137.30	137.77
137.40	137.85
137.50	

For additional frequencies search between 136.00 and 137.00.

Monitoring military satellites

Military satellites can also be monitored. The U.S. Navy FLEETSATCOM transmits between 261.00 and 263.00 MHz (FM mode), and frequency spacing is 25 kHz. Although most communications are scrambled, there is clear voice to be heard.

Satellite communications are very weak, and usually cannot be monitored without an outside antenna. Pre-amps are also utilized to boost the signal on specific frequencies. If you're interested in satellite monitoring, there are several books dedicated to the subject. Visit your local library or ham radio outlet store for more information.

Police frequencies

The nationwide police emergency frequency is 155.475 MHz. The frequency isn't active in every state, but it is expected to eventually provide a radio link among all fifty states. Additional police frequencies are shown in Table 6-23. Tactical frequencies, often used by SWAT and stakeout teams, may not be listed in local directories. Unpublished frequencies can be located by searching the range of a known frequency. Here's an example. Suppose that your local police are operating on 453.30 MHz. To find additional frequencies, program your scanner radio to search between 453.00 and 454.00 MHz. The procedure works especially well if several listening sessions are scheduled at various intervals. Don't listen at the same time every day. Some frequencies will be more active at night, while others may be active during the morning. It's a high-tech game of hide and seek that requires skill and patience.

**Table 6-23. A sampling
of common police frequencies**

Frequency
42.02 to 42.98
44.62 to 46.02
154.65 to 156.21
159.09 to 159.21
453.0125 to 453.9625
460.0125 to 460.5625
810.00 to 816.00
855.00 to 861.00

Note: The frequency ranges provided are not complete. There are hundreds of additional frequencies that may or may not be used by your local police.

Sports teams

If you've never taken a scanner radio to a sports event, you're in for quite a surprise. A few of the popular sports frequencies are presented in Table 6-24. Your scanner radio can place you in the driver's seat of a race car or on the sidelines of a pro football game. And as you have already learned, it's also possible to monitor the TV and radio crews that are reporting on the event.

Table 6-24. Sports frequencies

Frequencies
Bicycle races—154.600
Daytona—464.55
NASCAR auto races—464.50, 469.50
National Rifle Assoc.—467.60
Sports Car Club of America—452.94, 456.388
National Football League—461.238, 461.312, 461.338, 462.812

If you attend a sports event, don't forget to bring along an extra set of batteries. Battery packs are almost guaranteed to die at the height of the action! A comfortable earpiece is another item that will help to extend battery life. Listening through the speaker uses more power and may also disturb your neighbors.

More adventures

The scanning adventures in this chapter are only a very small sampling of what you can monitor. With a scanner radio at your side, you can explore adventures on the ground, in the air, or on the high seas. And as you have already learned, it's also possible to explore radio communications from outer space. Best of all, you can hear almost everything from the safety and comfort of your favorite chair!

7
CHAPTER

Finding the action

Frequency listings

Hurry! What are the Space Shuttle frequencies? If you're a typical hobbyist, you'll spend several minutes looking through a stack of notes to find the answer. When you need to find a specific frequency, there usually isn't time for detective work. If you can't place your fingers on the frequency within a few seconds, you'll probably miss the best part of the action.

To solve the problem, we've organized this chapter into an alphabetical listing of the nationwide frequencies. Need the FBI frequencies? Simply turn to F listings. Can't remember the baby monitor frequencies? You'll find them under the B listings. Finding a frequency has never been easier. If you know the name of the agency, it only takes a few moments to locate specific frequencies. All frequencies are in megahertz and mode changes and/or special requirements are listed with the respective frequency.

Aeronautical

Civilian

Arrivals and departures 123.60 to 123.650 MHz (search between frequencies)

Air traffic control 118.0 to 121.4, 123.675 to 128.80, and 132.025 to 135.975 MHz (search between frequencies)

Emergency 121.50 and 122.0 MHz

Flight service 121.975 to 122.675 MHz (search between frequencies)

Flight schools 123.50 MHz

Flight tests 123.525 to 123.575 MHz (search between frequencies)

Goodyear blimp 123.05, 151.625, 465.9125, 465.9375, and 465.9625 MHz

Ground support vehicles 121.6 to 121.95 MHz (search between frequencies)

Helicopters 122.975 to 123.075 MHz (search between frequencies)
Pilot requests for repairs 129.30, 129.70, 130.25, 130.60, 130.65, 130.85, 131.15, 131.425, and 132.00 MHz
Pilots talking between planes 123.45 MHz
Search and rescue 123.10 to 123.125 MHz (search between frequencies)

Military

Air Force One 407.85 and 415.70 MHz
Military control towers 236.60 MHz
Emergency 243.00 MHz
Flight service stations 255.40 MHz
FAA/Civilian tower 257.80 MHz
Naval aircraft 277.80 MHz
Search and Rescue 282.80 MHz
Strategic Air Command 311.00 MHz
Naval control towers 340.20 MHz
Pilot to dispatch 372.20 MHz
Air Refueling 239.80, 255.40, 255.60, 259.30, 273.50, 289.40, 290.90, 311.00, 319.80, 321.00, 324.30, 338.60, 364.20, 381.30, and 395.90 MHz

Air shows

Control tower frequencies 123.40 and 123.45 MHz
Blue Angels 142.0, 143.0, 241.40, 251.60, 275.35, 384.40
Thunderbirds 120.45, 126.20, 141.85, 413.025

Aircraft phone calls
(Calls made from air to ground)
459.70 and 459.975 MHz. Also search between 894.00 to 896.00 MHz in AM mode.

Alcohol, Tobacco, and Firearms

165.2875, 166.5375, 166.4625, 165.9125, 173.8875, and 168.00 MHz

Amateur radio (ham)

Search between:
28.00 to 29.70; 50.00 to 54.00; 144.00 to 148.00; 222.0 to 225.0; 420.00 to 450.00; 902.00 to 928.00; 1240.00 to 1300.00 MHz.

American Red Cross

47.42, 47.46, 47.50, and 47.65 MHz. Also search between 856.00 and 866.00 MHz.

Baby monitors

49.830, 49.845, 49.86, 49.875, and 49.89 MHz

Border Patrol

408.20, 408.225, 408.275, 413.65, 413.775, and 417.05 MHz

Bugs and body microphones

149.35, 165.912, 171.45, and 172.00 MHz

Bureau of Mines

34.85, 36.17, 40.60, 167.95, 168.55, 414.85, and 417.575 MHz

CB radio

Channel	1	26.965	Channel	2	26.975
	3	26.985		4	27.005
	5	27.015		6	27.025
	7	27.035		8	27.055
	9	27.065		10	27.075
	11	27.085		12	27.105
	13	27.115		14	27.125
	15	27.135		16	27.155
	17	27.165		18	27.175
	19	27.185		20	27.205
	21	27.215		22	27.225
	23	27.255		24	27.235
	25	27.245		26	27.265
	27	27.275		28	27.285
	29	27.295		30	27.305
	31	27.316		32	27.326
	33	27.336		34	27.346
	35	27.355		36	27.365
	37	27.376		38	27.385
	39	27.395		40	27.406

Cellular phones

Search 823.00 to 851.00 and 870.00 to 896.00 MHz*

Coast Guard (U.S.)

156.80, 157.05, 157.10, 157.15, 157.175, and 282.80 MHz

Cordless phones

46.61, 46.63, 46.67, 46.71, 46.73, 46.77, 46.83, 46.87, 46.93, and 46.97

Customs Service (U.S.)

162.825, 165.2375, 166.4375, 166.4625, 165.7375, 165.4625, and 166.5875 MHz.

*The Electronic Communication Privacy Act (ECPA) prohibits cellular phone monitoring.

Federal Bureau of Investigation (FBI)

165.90, 166.825, 167.2375, 167.2625, 167.4375, 167.50, 167.525, 167.5625, 167.5875, 167.6125, 168.85, 172.475, and 173.10 MHz

Federal Communications Commission (FCC)

41.060, 167.05, 172.050, and 172.80 MHz.

Federal Emergency Management Agency (FEMA)

138.225, 138.575, 139.10, 139.825, 139.45, 139.225, 139.95, and 140.025 MHz.

Federal prisons

170.875, 170.925, and 170.065 MHz.

Federal protection Service

415.20, and 417.20 MHz.

Fish & Wildlife Service (Dept. of Interior)

34.40, 34.45, 34.80, 34.85, 40.40, 408.675, 410.625 MHz.

Golf (PGA)

464.50 MHz.

Hospital and ambulance

155.34, 155.28, 155.16, and 155.22.

Med channel	Base	Mobile
1	463.00	468.00 MHz
2	463.025	468.025 MHz
3	463.05	468.05 MHz
4	463.075	468.075 MHz
5	463.10	468.10 MHz
6	463.125	468.125 MHz
7	463.15	468.15 MHz
8	463.175	468.175 MHz
9	462.95	Dispatch MHz
10	462.975	Dispatch MHz

Internal Revenue Service (IRS)

165.95, 167.00, 166.4625, 166.00, 418.225, and 418.175 MHz

Itinerant frequencies

27.49, 35.04, 43.04, 151.505, 151.625, 158.40, 451.80, 456.80, 464.50, and 464.55 MHz.

Maritime

Channel	Frequency (in MHz)	Channel	Frequency (in MHz)
1	156.05	60	156.025 and 160.625
2	156.10 and 160.70	61	156.075 and 160.675
3	156.15 and 160.75	62	156.125 and 160.725
4	156.20 and 160.80	63	156.175 and 160.775
5	156.25 and 160.85	64	156.225 and 160.825
6	156.30	65	156.275 and 160.875
7	156.35	66	156.325 and 160.925
8	156.40	67	156.375
9	156.45	68	156.425
10	156.50	69	156.475
11	156.55	70	156.525
12	156.60	71	156.575
13	156.65	72	156.625
14	156.70	73	156.675
15	156.75	74	156.725
16	156.80	75	not assigned
17	156.85	76	not assigned
18	156.90	77	156.875
19	156.95	78	156.925 and 161.525
20	157.00 and 161.60	79	156.975 and 161.575
21	157.05	80	157.025 and 161.625
22	157.10	82	157.075 and 161.675
23	157.15	83	157.175 and 161.775
24	157.20 and 161.80	84	157.225 and 161.825
25	157.25 and 161.85	85	157.275 and 161.875
26	157.30 and 161.90	86	157.325 and 161.925
27	157.35 and 161.95	87	157.375 and 161.975
28	157.40 and 162.00	88	157.425 and 162.025

McDonald's order window

35.02, 154.57, and 154.60 MHz.

Nationwide police emergency

155.475 MHz (Not active in all areas.)

Navy (U.S)

138.325, 138.65, 138.85, 138.975, 140.575, 140.82, 148.83, 140.895, and 160.375, MHz.

Nuclear search teams

167.95, 167.85, 169.60, 172.30, 164.10, 164.225, and 167.825 MHz.

Nuclear Regulatory Commission (NRC)

167.875, 168.45, 169.10, and 172.75 MHz.

Park Service, national

151.445, 164.425, 164.475, 166.95, 171.625, and 171.725 MHz.

Postal Service

164.70, 166.275, 169.00, 409.225, 412.275, 418.30, 169.85, 171.175, 408.00, 408.05, 409.10, 409.225, and 409.525 MHz.

Racing

Dale Earnhardt 469.0125 and 464.0125 MHz
Ernie Irvan 464.30 MHz
Bill Elliott 853.50 and 853.4875 MHz
Michael Waltrip 466.30 and 461.30 MHz
Richard Petty 865.665
Other racing frequencies 451.90, 461.15, 464.30, 466.65, 468.5125, 468.9375, 469.50, 851.50, 851.575, and 856.500 MHz.

Railroads

159.81, 159.93, 160.05, 160.185, 160.20, 160.215, 160.23, 160.245, 160.26, 160.275, 160.29, 160.305, 160.32, 160.335, 160.35, 160.365, 160.38, 160.395, 160.41, 160.425, and 160.44 MHz.

Red Cross

47.42 MHz.

Russian orbiting spacecraft

143.625 and 166.135 MHz.

Satellites

Search between 135.575 and 135.625 MHz.

Secret Service

164.65, 164.80, 164.8875, 165.215, 165.375, 165.5125, 165.7875, 166.5125, 166.70, and 167.025 Mhz

Space Shuttle

259.70 and 296.80 (AM mode) MHz

Television crews

Search between 450.05 to 450.925, and 455.05 to 455.925 MHz. Also search between 942.00 to 952.00 MHz.

Transportation Safety Board
165.7625 MHz

Treasury Department
166.4625, 165.205, 165.215, and 165.4875 MHz.

Weather Service (N.O.A.A.)
Search between 162.40 and 162.55 MHz.

White House communications
167.7875, 165.375, 169.925, 407.925, 165.213, 165.025, 164.8875, 164.40, 166.70, 165.65, 164.10, 166.4625, 162.6875, and 171.2875 MHz.

Wireless microphones
33.40, 169.445, 169.505, 170.245, 170.305, 171.045, 171.105, 171.845, and 171.905 MHz.

Frequency hunting

Although we have presented hundreds of frequencies that are active nationally, there are thousands of local frequencies that can also be explored. In every large city, you'll be able to monitor utility companies, shopping malls, institutes, museums, zoos, commuter buses, trains, and many additional agencies. Discovering the hidden frequencies in your city is one of the many exciting aspects of scanning.

Where to search

Searching for frequencies, however, can be a frustrating experience. It might take hours of intense monitoring to identify newly discovered frequencies. To reduce your search time, we recommend using the *Consolidated Frequency List*, in *Police Call* (Table 7-1). The list is arranged numerically and it identifies the user of individual frequencies. Suppose that you're looking for the frequencies used by the U.S. Postal Service. As you can see, the postal service is abbreviated UPO. Simply run your finger down the frequency list and note each frequency annotated with UPO. Frequencies for other agencies can be located in the same manner. After the appropriate frequencies are selected, punch them into your scanner and keep the ones that are active.

Table 7-1. Consolidated frequency guide

Frequencies	Abbreviations	Key	
169.00	UIL, UPO, UVA	BIFC	Fire cache
169.025	UCE, UIL	UCE	Environmental
169.0375	UDC	UCW	Weather Service
169.05	USD	UCM	Maritime

Table 7-1. Continued.

Frequencies	Abbreviations	Key	
169.075	UCM, UCW, UIL	UDC	Govt. Dist. of Columbia
169.10	UGF, UIR, UNR	UGF	Forest Service
169.125	UGF, UIP, UVA	UIL	Land Management Bureau
169.150	UGF, BIFC	UIP	National Parks
169.175	UGF	UIR	Reclamation Dept.
169.20	UDC, BIFC	UNR	Nuclear Regulatory Comm.
		UPO	Postal Service
		UVA	Veterans Affairs

Source: Adapted from Police Call.

Name that frequency

Think quickly! What are the space shuttle frequencies? If you had to go back to the alphabetical listings, don't get discouraged. In a few months, certain frequencies will become memorized and you'll recall them with ease. For now, simply place this book near your scanner radio and use it as a handy reference guide.

8
CHAPTER

Computers and scanning

Cellular car phones are an example of computer-controlled radio communications. When you press the "send" button on your cellular phone, a radio transmission is sent to the nearest cellular antenna tower (Fig. 8-1). A computer converts the radio signal into an electrical signal that is then sent over conventional phone lines. To receive a cellular phone call, the computer must reverse the procedure. A land-line phone call is converted into a radio signal and transmitted to your cellular phone.

8-1
Cellular phone antenna tower.

Computers are also responsible for tracking the position of your cellular phone. As you travel between individual antenna sites, the computer automatically executes a *cellular hand-off* (Fig. 8-2). The technology allows your airborne cellular phone signal to be transferred to the nearest antenna tower. The hand-off is accomplished with lightning speed and cannot be detected by the cellular customer.

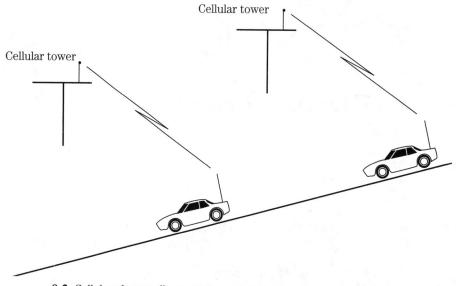

8-2 Cellular phone calls are automatically switched between towers.

Cellular phone frequencies

Cellular phones can operate on 15 different frequencies. A computer is used to select an unused frequency and to randomly change the frequency at various intervals. During a five-minute cellular conversation, the frequency might change several times. The frequency changes are timed perfectly and cannot be detected by the customer.

The abrupt frequency changes can, however, be detected by third party listeners. The monitored cellular conversation will suddenly change from one frequency and will appear on another. If you monitor between 870 and 890 MHz, you'll hear the frequency changes that occur during a cellular conversation. Remember, however, that cellular phone monitoring is illegal. See chapter 5 for additional details.

Trunking

Trunking is another type of radio communication that is computer controlled. In a trunked system, it is not uncommon for the police, fire, ambulance, highway department, and dog catcher, to be assigned to the same group of frequencies. However, no individual or group is assigned to a single frequency.

When the dog catcher presses the microphone button, a computer automatically selects an unused frequency. When a police officer pushes the mic button, the computer selects the next unused frequency. Trunked systems usually contain a group of 20 frequencies. There are a limited number of trunked systems that operate with as few as five frequencies.

Trunk system monitoring

The computer's ability to randomly select and assign specific frequencies has prompted news reporters to falsely claim that trunked radio systems cannot be monitored. The complexity of the trunked system has also frustrated scanning hobbyists. Because there are no assigned frequencies, hobbyists complain that it's impossible to hear a complete conversation. One minute you're listening to the dog catcher, a few seconds later, you're listening to the street cleaner. Can trunked systems be monitored? Can you follow the conversation as it skips across 20 frequencies or more? The answer to these questions is a resounding yes!

Trunk busting

In a trunked system there are ten channels (or more) that may be assigned to a variety of agencies (Fig. 8-3). Similar to cellular phones, the programming and switching of frequencies is accomplished by computers. Unlike cellular phones, trunked systems will not switch to another frequency in the middle of a conversation.

When a mobile user presses the microphone button, the vehicle's radio automatically tunes to an empty frequency. At the repeater site, a computer tunes the base radio to the same frequency as the mobile unit. As soon as the microphone button is released, the computer assigns the user to a new frequency.

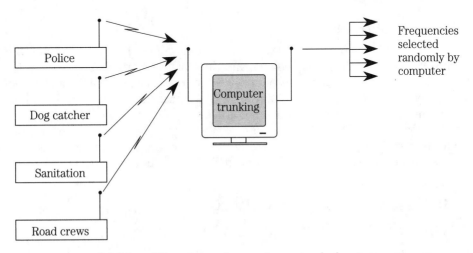

8-3 Many different agencies can share a trunked system.

Signaling and data frequencies

On some systems, as many as four frequencies may be used for signaling and data transfer. Short bursts of encoded information are constantly sent between the main computer and the patrol vehicles. The information is used by the computer to determine the status of the patrol car and to provide the base dispatcher with the vehicle's identification number. In an emergency, an officer can simply press a button and the dispatcher will know the ID of the unit requiring assistance.

The signaling and data frequencies will produce a loud, buzzing sound in your scanner radio's speaker. The first step to monitoring a trunked system is to lock out all of the signaling and data frequencies. The remaining frequencies, that are not used for signaling and data purposes, are conversation frequencies that can be monitored. However, be advised that both the signaling and data channels will change from day to day. To be successful, the hobbyists must determine the active voice and data frequencies on a daily basis.

Following the action

The delay feature on your scanner radio should be deactivated. A trunked system will change frequencies as soon as the user releases the microphone button, and use of the delay feature will not permit the scanner radio to instantly follow the conversation to the next frequency.

Repeated pressing of the scan button might be necessary to follow a specific conversation. The first few words of the conversation might be lost, but that won't prevent you from understanding the communication. Utilizing two scanner radios is another method that has proven successful. One scanner radio is programmed with frequencies in ascending order, the other programmed in descending order. If scanner #1 skips part of the action, scanner #2 will provide the missing portion. At first it might be a little confusing, but your ears will quickly become accustomed to listening to several scanner radios.

The world of trunking

Each manufacturer's trunked system is different. A few of the leading manufacturers are Motorola, Johnson, Midland, and Uniden. The Motorola system was described previously. In the Johnson trunked system, both signaling and voice frequencies are transmitted simultaneously. The signaling frequency range is just below the speech frequency. However, don't become discouraged. All the systems can be monitored by using the procedures that were previously outlined. As you become more experienced, try to discover the type of system that is in use. It then becomes fairly easy to tailor your monitoring techniques to fit the mold.

Continuous Tone Coded Squelch System (CTCSS)

The radio spectrum does not contain a limitless supply of frequencies. As you have already learned, there are countless individuals and agencies that utilize radio communications on a daily basis. Frequency availability is limited and expensive. The cellular industry, for example, paid millions of dollars to acquire the frequencies used by cellular phones. Corporations that cannot afford to purchase a specific group of frequencies are forced to operate within a limited and often crowded band of frequencies.

To solve the overcrowding problem, the Continuous Tone Coded Squelch System (CTCSS) was designed to allow several agencies or businesses to use the same frequency. A CTCSS-capable radio simultaneously transmits a subaudible tone with every voice message. The intended receiver recognizes the tone and allows the transmitted message to be heard. Anyone transmitting with the improper tone will not be heard. With CTCSS technology, a construction firm can separate management and labor communications. The company president can transmit a radio message that will only be heard by a select group of supervisors. Truck drivers, on the same frequency (Fig. 8-4), would not hear management communications.

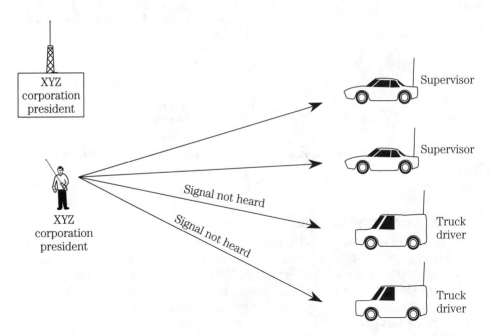

8-4 CTCSS tones allow multiple users to transmit and receive on a single frequency.

A familiar example

The most familiar example of CTCSS technology can probably be found in your local fire department. A single alerting frequency is monitored by dozens of individual fire companies (Fig. 8-5). Because each fire company is assigned a specific tone, the dispatcher can select the fire companies that will respond to the fire. The remaining fire companies that monitor the same frequency will not be disturbed.

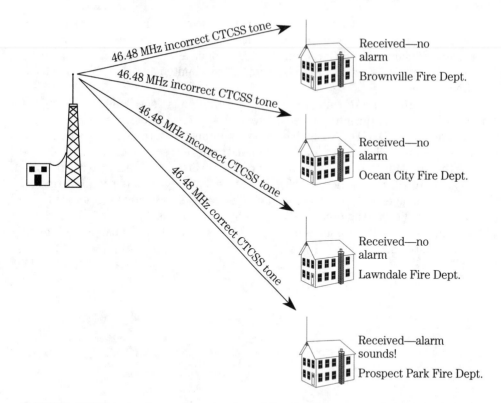

8-5 With CTCSS technology, a large number of fire departments can monitor a single frequency.

Radio tones in your home

The use of tones to call individuals can also be found within your home. A touch-tone telephone (Fig. 8-6) uses tones to dial numbers. The most widely used method is known as *dual tone, multi-frequency (DTMF)*. Each press of the button on your phone sends two audio frequencies. Filters in the phone company's equipment can recognize the tones and activate the appropriate circuits. The tones used by CTCSS perform the identical function. A specific group of tones are used to activate a single receiver or group of receivers.

Standard tones

The 32 standard CTCSS tones can be seen in Table 8-1. In commercial equipment the tones are filtered out, the user can't hear them. If you monitor the tones you'll hear a soft buzzing sound. There are a few CTCSS-capable scanner radios on the market. The Uniden Bearcat series can be programmed to respond to CTCSS tones. The feature will allow you to focus your monitoring on a specific user. A few commercial frequency guides are beginning to include CTCSS tone frequencies for a variety of agencies.

American Telegraph and Telephone (AT&T)

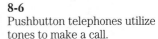

8-6
Pushbutton telephones utilize tones to make a call.

Table 8-1. Standard CTCSS tones

Frequencies in Hz					
67.0	85.4	107.2	131.8	162.2	203.5
71.9	88.5	110.9	136.5	167.9	210.7
74.4	91.5	114.8	141.3	173.8	218.1
77.0	94.8	118.8	146.2	179.9	225.7
79.7	100.0	123.0	151.4	186.2	233.6
82.5	103.5	127.3	156.7	192.8	241.8

Computer-Aided Dispatching (CAD)

Computer-Aided Dispatching (CAD) is used to consolidate police, fire, and medical dispatching into one location. When a complaint or call for help is received, the information is entered into a computer. The required information is limited to the caller's name, location, and nature of the incident. After the call is received, the computer requires less than thirty seconds to sort the information and send it to the appropriate dispatcher. In a large city, the inside of a CAD communications center might resemble the drawing in Fig. 8-7. Incoming calls are received by the complaint taker, and the computer is used to route the call to one of several dispatchers.

Vehicles can also be equipped with CAD technology. A small box that contains several buttons is attached to the mobile radio. Pushing one of the buttons sends a digital burst of information between the vehicle and the dispatch center. The information usually consists of routine reports: "Arriving at scene," "out of service," "in service," "help needed," etc.

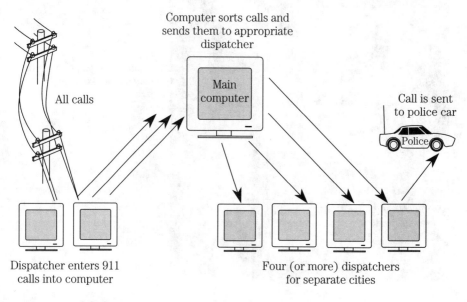

8-7 Typical layout of an emergency communications center.

Mobile Data Terminals (MDTs)

The most sophisticated form of mobile communications are called *Mobile Data Terminals (MDT)*. Mobile MDTs are miniature computers with a keyboard and monitor screen. A police officer can use an MDT to instantly access license information, vehicle registration, and possible wants and warrants on the driver. The information is received without the need for voice communications.

At this point, I know what you're thinking. Can MDTs be monitored with a scanner radio? MDTs convert the data into numeric values that are transmitted in short bursts. The data is sent over a separate frequency that you can receive, but no voice will be heard. The sound of a data frequency includes high-pitched chirps and squeals that are very annoying to the listener.

Trading frequencies

Frequencies are the baseball cards of scanning. Hobbyists collect, trade, and sell frequencies on a regular basis. Even if you're new to the hobby, it won't take very long for you to collect several hundred frequencies. Seasoned scanning hobbyists usually maintain frequency lists that contain thousands of local and national scanning frequencies.

Frequencies can also be purchased commercially. There are several companies that sell local and national frequency lists on micro film and/or computer disks (Fig. 8-8). See chapter 10 for a list of frequency suppliers.

8-8
Frequencies can be purchased through the mail.

Computerized scanning

The difficult task of collecting, sorting, and printing large groups of frequencies is made possible by computers. Frequency management programs are available on computer disks and are sold commercially. The programs (software) allow the user to add, delete, and sort frequencies numerically and alphabetically by agency. Scanning software that actually connects your computer and scanner radio together is also available through mailorder catalogs. With this type of system, a computer automatically logs all the frequencies that are received by the scanner.

Futuristic scanning

As scanner radios and scanning accessories become more sophisticated, the ability to monitor the airways will continue to expand. In the future, enhanced listening abilities will undoubtedly take us on new adventures. To join the fun, all you need is a scanner radio. Simply press a few buttons and your adventures will never end.

9
CHAPTER

Establishing a listening post

A listening post can be placed in practically any location. Hobbyists have placed impressive listening posts in attics, basements, and closets. If your free space is limited, a folding table, located in the corner of a room, is perfectly acceptable. If you're thinking about establishing a specific monitoring area, here are several hints and ideas to consider.

Locations

Attic monitoring

A finished attic is an ideal place for a listening post. The close proximity to the roof will limit coax cable lengths, reduce signal loss, and provide privacy. And don't forget that scanning antennas can also be installed in the attic (see chapter 2). If your attic hasn't been converted into a living area, this would be the perfect opportunity to shape your attic into a custom listening area.

Basement monitoring

Basements that are dry and free from dust are ideal locations for a listening post. The two main concerns associated with basement monitoring are: (1) below ground reception demands the use of outdoor antennas, and (2) excessively long cable runs between the outside antenna and basement.

Monitoring below ground with an inside antenna probably won't work very well. You'll need to install an outside antenna and route the coax cable into the basement. It is recommended that the coax cable length be limited to 100 feet or less. Cable runs in excess of 100 feet will require a signal amplifier.

Window locations

As you examine your home for possible listening areas, it's important to remember that window locations provide easy access for coax cables. Windows, however, can also be the source of several problems. Noninsulated windows can "sweat" and drip moisture onto your expensive equipment. Windows can also produce drafts. During the winter months, a cold and drafty location can make monitoring very uncomfortable.

Power requirements

Your listening post will require more than one power outlet. Since scanning equipment doesn't draw large amounts of current, multiple outlet power strips, like the one shown in Fig. 9-1, are used in practically every listening post. Power outlets are also available with on/off switches and fused protection. Regardless of the type that is purchased, don't forget to look for the Underwriters Laboratories (UL) label.

9-1 Power outlet strips are popular problem solvers.

Surges and spikes

The electricity that passes through your house wiring is maintained at a constant voltage. Small kitchen appliances operate on 115 V. Larger appliances, electric dryers, and air conditioners require a constant voltage of 210 V. Inadequate wiring or overloaded circuits can cause a voltage surge. The on/off cycling of an air conditioning compressor can cause room lights to momentarily dim. When the lights return to their standard brightness, a voltage surge occurs. The microcomputer chips used in electronic gear can be severely damaged by a voltage surge.

Voltage surges can also be caused by factors that are many miles from your home. When a large factory or shopping mall suddenly shuts down for the evening, an increase in line voltage can travel through power lines and enter your home.

Lightning

Lightning is another uncontrollable factor that can cause a voltage surge. If lightning strikes a power line that is miles away from your home, the electrical surge, traveling at the speed of light, will reach your location in seconds. To protect your equipment from a voltage surge there are a variety of plug-in surge protectors (Fig. 9-2) on the market. Electrical power strips, that were mentioned earlier, can also be purchased with internal surge protectors (Fig. 9-3).

9-2
Plug-in surge protectors are
widely available.

9-3 Power strips are available with internal surge protectors.

Reusable protection

The majority of surge protection devices on the market cannot be reset. If a voltage surge is detected, the circuitry trips and the entire unit must be replaced. The more expensive units contain a reset button that reactivates the protective circuitry.

If your budget can handle the initial expense, it's more economical and practical to purchase a surge protector that can be reset.

Sitting comfortably

Selecting a quality chair for your listening post is a serious concern. A good quality, well-constructed chair is probably the most important accessory in your listening post. A comfortable chair that supports your back is an absolute necessity.

Depending on the layout of your listening post, a swivel chair with adjustable height setting, will allow you to turn and reach your equipment. Armrests are another option to consider, but they do require more space. Sitting areas with limited space will require the purchase of a chair that meets specific measurements. Again, don't overlook the importance of a comfortable chair. Take your time and select a chair that provides comfort without busting your budget.

Tape recording

Saving your scanning adventures with a tape recorder will provide you with lasting souvenirs. Conventional tape recording devices require the user to manually start and stop the recording, and tape machines that are simply allowed to run will produce long periods of silence and wasted tape. To solve the problem, a *voice activated* recorder can be plugged directly into the scanner's headphone jack. The recorder starts and stops the tape at the beginning and end of each conversation. Unattended voice activated recordings can be made while you're at work or home sleeping. A full eight hours of monitoring will usually fit onto a 60-minute cassette tape.

Readers who may already own a good quality tape recorder will not need to purchase a voice activated unit. Several companies make separate, voice activated controllers (Fig. 9-4), that can be installed between your tape recorder and scanner radio. The device will start and stop the tape recorder in unison with voice communications.

9-4
Voice-activated controllers link your scanner radio to your tape recorder.

Timer control

Appliance timers, like the one shown in Fig. 9-5, can be used to activate your monitoring equipment. Suppose that you want to monitor a group of frequencies between the hours of 1 and 4 AM. Simply plug your equipment into the timer and forget it. Depending on the timer that is purchased, it might be possible to select several on and off times within a 24-hour period.

9-5
Appliance timers can be used to control your equipment.

Cords and cables

Power cords should be mounted away from antenna cables (Fig. 9-6). Radio Shack has a variety of stick-on holders, wire ties, and other accessories (Fig. 9-7), that can be used for this purpose. Power cords and coaxial cables should also be labeled as shown in Fig. 9-8. The separation between the wire cable provides a margin of safety. Problems that might occur in the power cord and/or coaxial cable can be isolated and easily corrected.

Extension cords

Extension cords provide an easy solution to a variety of power related problems. Is the power cord too short? Need another outlet? No problem, reach for an extension cord. Don't settle for a small diameter, two-wire extension cord, similar to the one shown in Fig. 9-9. These types of extension cords are not designed to carry the increased power demands of your monitoring equipment. A three-wire, grounded extension cord, similar to the one shown in Fig. 9-10, is strongly recommended. The heavier gauge wire and grounded plug can easily carry the power requirements of your scanning gear.

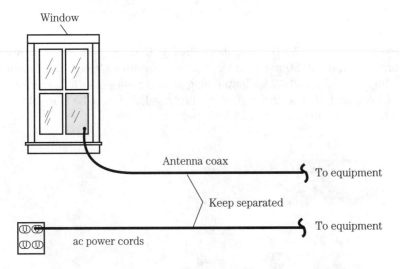

Window

Antenna coax

To equipment

Keep separated

To equipment

ac power cords

9-6 Power cords and coaxial cable should be separated.

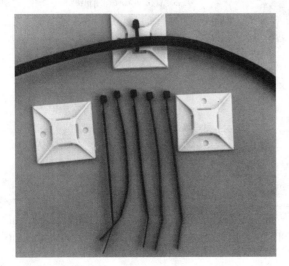

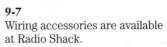

9-7
Wiring accessories are available
at Radio Shack.

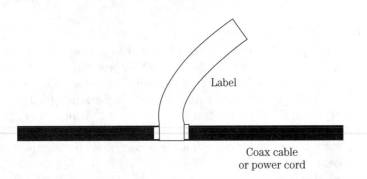

Label

Coax cable
or power cord

9-8 Stick-on labels help to identify power cords and coaxial cables.

9-9 Small diameter two-wire extension cords can be dangerous.

9-10
A heavy duty extension cord, with a grounded plug, is the safe way to add flexibility to a wall socket.

Extension cords should not be concealed under any type of carpet or rug. Walking on an extension cord can eventually crush the wires and short circuit the cord. Damage to your equipment and the possibility of a fire are common hazards that occur when extension cords are hidden under high-traffic areas. Extension cords should be placed in open areas and inspected regularly for signs of wear.

Line noise

The electrical circuits in your home are separated and controlled by circuit breakers (Fig. 9-11). Your kitchen outlets, for example, are usually isolated from the outlets in your living area (Fig. 9-12). The lighting circuits in most homes are also separate from the wall outlets (Fig. 9-13).

In older homes, the circuitry isn't so easily defined. A previous home owner may have added a baseboard wall socket and spliced the wires into the ceiling lights. If you plug your scanning equipment into a socket that is connected to fluorescent ceiling lights, electrical line noise can be heard on the scanner's speaker.

Fortunately, it's easy to isolate electrical circuits. Open your circuit breakers one at a time and check what appliances and/or lights don't work. Ideally, your scanning equipment should be plugged into an electrical circuit that does not include kitchen appliances, fluorescent lights, or devices that are controlled by the closing and opening of electrical contacts. Circuit alterations, if required, should be referred to a licensed electrician.

9-11
Typical circuit breaker panel.

Useful gadgets

White colored plastic, cut into small panels (Fig. 9-14), is ideally suited for recording your scanning frequencies and notes. Mount the panel on a wall and use dry erase markers to record newly discovered frequencies, comments, and other related scanning information (Fig. 9-15). After transferring the notes to a permanent log, the information on the plastic can be wiped away with a dry cloth.

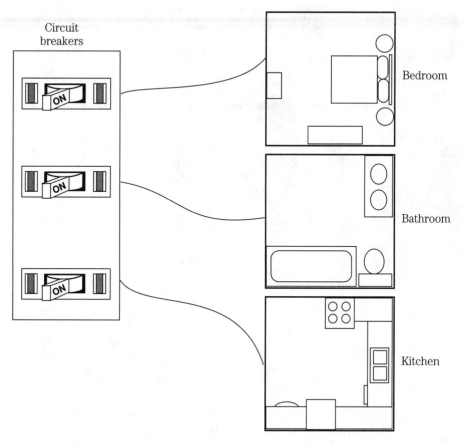

9-12 The circuits in your home are isolated for safety.

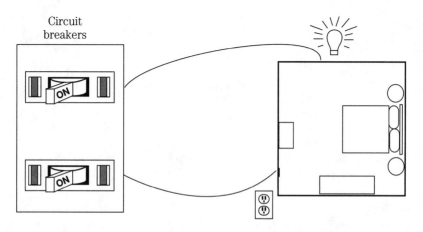

9-13 Lighting and wall outlets are usually separated.

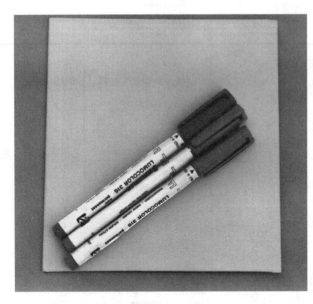

9-14
Colored plastic can be used to enhance your listening area.

New Frequency — 54.90

118.50 — AIRCRAFT

46.61 — ?

9-15 Writing on plastic panels is easy with dry-erase markers.

Small sections of white plastic can often be obtained from sign making shops. Excess material that cannot be utilized by the sign maker might be offered free of charge. Custom lengths of colored plastic can also be ordered from glass supply and repair shops. Check the yellow pages under glass or plastic.

Road maps

Nearly every listening post contains a road map. During an emergency or high-speed pursuit, a local road map will allow you to locate and follow the action. Local maps can be obtained from police or fire departments, local libraries, and department stores.

Using a road map in your listening post can be difficult, if not impossible. A fully opened map can be cumbersome to handle and difficult to refold. To solve the problem, road maps can be glued onto window shades (Fig. 9-16). Window shades can be purchased in a variety of sizes and they can be custom cut to your specific measurements. Mounting a window shade is easy and several shades can be mounted together (Fig. 9-17). When you need to refer to a map, simply pull down a shade!

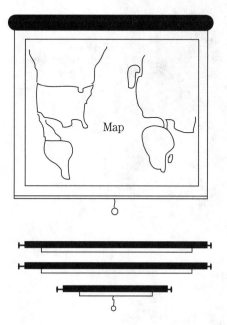

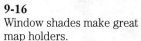

9-16
Window shades make great map holders.

9-17
Multiple shades can be mounted in small space.

Headphones

A quality set of headphones will prevent your listening activities from disturbing others. The headphones commonly used by scanner buffs can be seen in Fig. 9-18. These lightweight headphones are inexpensive and can be purchased in department stores.

The larger stereo headphones, shown in Fig. 9-19, are not recommended for scanning. Sure, they can be used, but radio communications will only be heard on one side of the headset. It might be possible, however, to manipulate the plug position to provide mono sound to both sides of the headset.

AM radio

Does your neighborhood have an all-news station on the standard AM broadcast band? If so, you can utilize an AM radio to find active frequencies. To monitor a major traffic delay, for example, you would punch in the police, ambulance, and traffic reporting frequencies. A problem at the airport would require you to activate your airport frequencies.

The all-news format of a local radio station can also test your scanning ability. As you become an experienced hobbyist, you should hear the action on your scanner radio long before it is announced on the news.

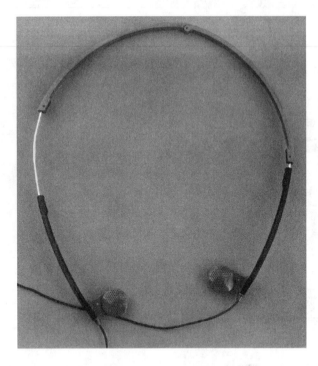

9-18
Lightweight headphones used
for private listening.

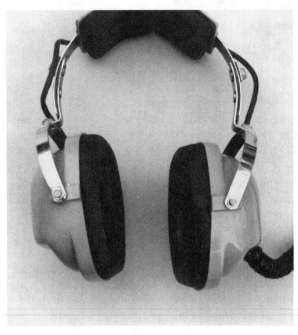

9-19
Headphones used for stereo
listening.

Frequency counters

Frequency counters are probably the most misused, and misunderstood, gadget on today's market. The frequency counter (Fig. 9-20) is a highly sensitive instrument that can capture and display transmitted frequencies. However, there are several limitations that must be understood. Frequency counters can only capture the radio signal frequency from transmitters that are within a few feet from your location. You can't, for example, use a frequency counter to catch frequencies from your favorite chair.

Frequency counters are typically used by scanner buffs to find the hidden frequencies of mobile and handheld transmitters. Again, you must be within close proximity to the transmitting antenna. A typical example would be a shopping mall security guard. As soon as the guard presses the microphone button on his handheld radio, your frequency counter will display the operating frequency. The information can then be entered into your scanner radio and you can listen to the action.

Your relative position to the transmitting antenna cannot be overemphasized. You must get as close as possible and you might need to adjust the position of your frequency counter by raising, lowering, or twisting your hand from right to left. It takes practice, skill, and a little luck to make it all work. If you understand the instrument's limitations, catching frequencies in this manner can be a lot of fun. For more information, contact OptoElectronics, 5821 NE 14th Ave., Ft. Lauderdale, FL 33334.

9-20
Frequency counter.

Spectrum display units

The *spectrum display unit (SDU),* or spectrum analyzer, (Fig. 9-21) will allow you to visually see the radio spectrum. Transmitted signals appear on the screen as a series of dots or solid lines. The dots or lines will produce peaks (Fig. 9-22) that can be instantly matched to specific frequencies.

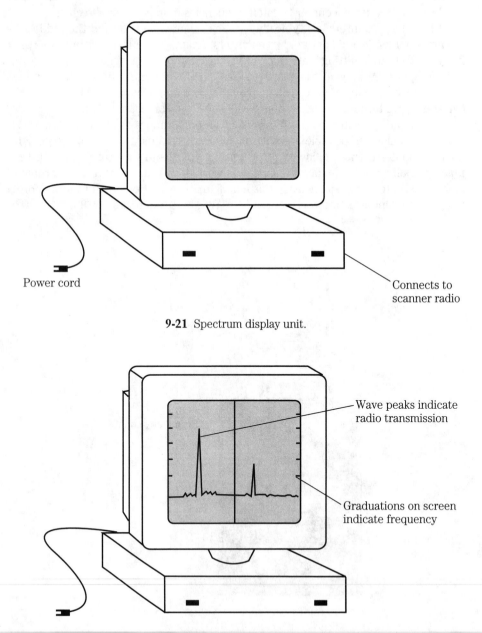

Power cord

Connects to
scanner radio

9-21 Spectrum display unit.

Wave peaks indicate
radio transmission

Graduations on screen
indicate frequency

9-22 Wave patterns visually identify specific frequencies.

The frequency coverage of spectrum display units is very large. You can visually examine and search for active frequencies within an entire frequency band. Unlike a scanner radio, which searches across individual frequencies, a spectrum analyzer allows you see an entire band and instantly zero in on active frequencies.

Spectrum analyzers can reduce your search for new frequencies to a mere fraction of the time normally required. Instead of waiting for your scanner radio to numerically stumble across an active frequency, you can use the analyzer to instantly see the transmitted signal from an unknown transmitter. For more information, contact Grove Enterprises, P.O. Box 98, Brasstown, NC 28902.

Scanner programming

The most highly organized, expensive, and modern listening post is useless if you don't know how to store the frequencies in your scanner radio. A scanner radio with only twenty channels isn't much of a problem. But as you already know, there are scanner radios that feature 400 channels and more. Programming 400 frequencies into one radio can be a lengthy and tedious task. Each frequency must be entered into a specific channel, one at a time. The first step in the process is to push your scanner radio aside and grab a pencil and paper.

Your scanner radio is separated into *banks* of frequencies. Each bank in your radio should be reflected on the paper. The numerical sequence of each bank will exactly match the display on your scanner radio (Fig. 9-23). After the banks are numbered, assign a title to each one. The titles that you choose will reflect your individual scanning interests. Your first bank will probably be titled *Police and Fire Frequencies*. The next step is to pencil in your local public safety frequencies (Fig. 9-24).

9-23
Scanner radio programming begins with a pencil and paper.

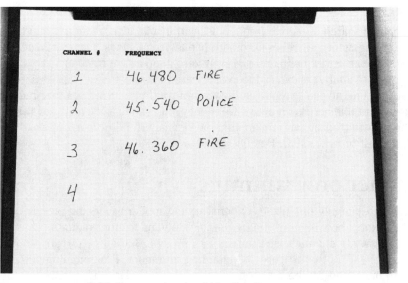

9-24 Frequencies should be listed on paper.

The second bank of frequencies should also be labeled to reflect your next area of interest. If you choose the *Civilian Air Band*, pencil in all of the aircraft frequencies that you want to monitor. Don't use a pen. Pencil marks are easy to erase and change. Continue naming the banks and adding the appropriate frequencies until all the banks are filled.

Transferring the frequencies from paper to your scanner radio is the final step. Take your time and check each entry. It is very frustrating to discover that you missed the action because of an incorrectly entered frequency. After all the frequencies are entered into your scanner, don't discard the paper list. Future deletions and changes should be made on the list and then transferred to your scanner radio. Maintaining a current list will allow you to see the programming at a glance. And should your scanner radio's memory fail, the list will be an invaluable reference source.

Scanning fun

Large, equipment-filled rooms are not required to enjoy the hobby of scanning. Expensive gear won't increase your level of satisfaction. A portable scanner radio, pushed into your coat pocket, can provide hours of enjoyable listening. As your interest and skill levels increase, new and more sophisticated equipment can be gradually purchased and added to your listening post. For now, stick to the basics and give yourself time to become familiar with your new hobby.

10
CHAPTER

Scanning clubs, books, and magazines

If you're living in or near a large city, there are hundreds, if not thousands, of radio frequencies that you can monitor. Locating active frequencies in your town may require hours of dedicated listening. To reduce your search time, it is recommended that you join a scanning club. Membership in a club will introduce you to scanning hobbyists with very diversified backgrounds. Retired members can provide you with frequency lists that have been confirmed by hours of dedicated listening. Members actively involved with computers can help you to computerize your listening post.

Scanning clubs are scattered throughout the country and their numbers are steadily growing. If you can't find a club that represents your listening area, contact the nearest club and ask them for assistance. New clubs are born every day, and it's quite possible that a new club may be forming near you. If a local club isn't available, don't overlook the possibility of starting your own club. Monthly meetings can be held in a private home or in a local restaurant. Scanning hobbyists in your area can be located by advertising in newspapers and magazines. Potential club members can also be located by placing messages on computer bulletin boards.

How to choose a club

As already mentioned, membership in a local club can provide you with an endless supply of local frequencies. Club membership also strengthens the hobby by providing hobbyists with a place to meet new friends and exchange ideas. To help you choose wisely, here are a few questions and comments that should be considered.

(1) Is the club a nonprofit organization? Nonprofit clubs are probably the best type to join. Each dollar that the club earns is used to improve and enhance the club's growth. (2) Are club bulletins published monthly or bi-monthly? Scanner buffs looking for a monthly bulletin may not be happy with a club that only publishes six issues per year. On the other hand, a monthly bulletin with limited material will not

be as informative as a well organized and action packed bi-monthly publication. (3) Are the club's publications dependent upon reader submissions? Club bulletins that are dependent on reader submissions can be a disappointment. If you joined the club because it included a column on scanning your neighborhood, a lack of reader input can prevent your favorite column from appearing in each publication. (4) How old is the club? The age of the club should be compared to the number of subscribers. The older, established clubs will naturally have impressive membership lists. Don't overlook the fact that club membership can also be influenced by geographical location. East and west coast clubs typically have large memberships. Clubs located in sparsely populated areas will have a much smaller number of subscribers. (5) Does the club have a beginner's column? It's frustrating to join a club that is light years ahead of your skill level. A well written beginner's column will allow you to gain years of valuable experience.

The following list represents a small sampling of scanning clubs that are scattered throughout the nation. Readers are encouraged to contact the clubs for additional information.

Radio clubs

California

Radio Communications Monitoring Association (RCMA)
P.O. Box 542
Silverado, CA 92676

Bay Area Scanner Enthusiasts (BASE)
4718 Meridian Ave., 265
San Jose, CA 95118

Southern California Area DXers (SCADS)
3809 Rose Ave.
Long Beach, CA 90807-4334

Canada

Association of Manitoba DX   uy ers (AMANDX)
30 Becontree Bay
Winnipeg, Manitoba R2N 2X9, Canada

Ontario DX Association (ODXA)
P.O. Box 161 Station A
Willowdale, Ontario M2N 5SB, Canada

Colorado

Rocky Mountain Monitoring Enthusiasts
11391 Main Range Trail
Littleton, CO 80127

Florida

Central Florida Listener's Club
956 Woodrose Court
Altamonte Springs, FL 32714

Illinois

Chicago Area Radio Monitoring Association (CARMA)
6536 N. Francisco 3E
Chicago, IL 60645

Massachusetts

Metro Radio Systems
P.O. Box 26
Newton Highlands, MA 02161

Maryland

Radio Monitors of Maryland
P.O. Box 394
Hampstead, MD 21074

Michigan

Michigan Area Radio Enthusiasts (MARE)
P.O. Box 81621
Rochester, MI 48308

New Jersey

Bayonne Emergency Radio Network (BERN)
P.O. Box 1203
Bayonne, NJ 07002

New York

Worldwide TV-FM DX Association (WTFDA)
P.O. Box 514
Buffalo, NY 14205

Long Island Sounds Monitor
2134 Decker Ave.
North Merric, NY 11566

NYC Radio
199 Barnard Ave.
Staten Island, NY 10307

North Carolina

Triangle Area Scanner Group
P.O. Box 28587
Raleigh, NC 27611

Ohio

Cincinnati Area Monitoring Exchange (MONIX)
7917 Third St.
West Chester, OH 45069-2212

Toledo Area Radio Enthusiasts (TARE)
6629 Sue Lane
Maumee, OH 43537

Pennsylvania

Susquehanna City Scanner Club
P.O. Box 23
Montrose, PA 18801

Texas

Houston Area Scanners & Monitoring Club
909 Michael
Alvin, TX 77511

Utah

Wasatch Scanner Club
2872 W. 7140 South
West Jordan, UT 84084-2917

Wisconsin

Wisconsin Area Monitoring Club
122 Greenbriar Drive
Sun Prairie, WI 53590

Additional club information

To obtain additional information on scanning clubs in your area, contact REACT International, P.O. Box 998, Wichita, Kansas 67201. REACT is a volunteer team that utilizes CB radio to provide safety communications for special events. REACT is a national organization with members in all fifty states. The REACT team in your area can help you to contact fellow scanning enthusiasts. For additional information, or to locate the REACT team near you, write to REACT at the above address.

Scanning Books

There are dozens of books that can help you to expand your knowledge of scanning. You can learn how to monitor satellites, modify your scanner radio, listen to railroad communications, and much more. If you can't find a specific book in your local library, don't get discouraged. Many of the books are self published, and are not available in libraries or book stores. All of the books listed below, however, can be obtained through mail-order catalogs. The popular mail-order suppliers are: Grove Enterprises, P.O. Box 98, Brasstown, NC 28902, and CRB Research, P.O. Box 56, Commack, NY 11725.

Books

Compendium of American Railroad Radio Frequencies by Gary L. Storm and Mark J. Landgraf

US Maritime Frequency Directory by Robert Gad and Robert Coburn

Receivers and Scanners Pricing Guide by Bob Grove

Tune In On Telephone Calls by Tom Kneitel

Monitoring the Military by Daryll Symington

Sports and Entertainment Frequency Directory by Bob Grove

Official Aeronautical Frequency Directory by Robert Coburn

Weather Satellite Handbook by Ralph E. Taggert

Receiving Antenna Handbook by Joe Carr

Scanner Modification Handbook, Vol. 1 & 2 by Bill Cheek

The Antenna Handbook by Clem Small

Magazines

The two top magazines in the scanning hobby are *Monitoring Times*, P.O. Box 98, Brasstown, NC 28902, and *Popular Communications*, 76 North Broadway, Hicksville, NY 11801. Both magazines publish 12 issues per year and feature columns on all aspects of listening to radio communications. Check your local news stand or contact the publisher at the address provided.

As your scanning interests expand, you might discover the need for a more technical and detailed explanation of specific topics. Antenna design, antenna theory, and coaxial cable impedance are just a few of the interesting areas that you can explore. Amateur radio operators (hams) can provide you with detailed explanations and hands-on help. The American Radio Relay League (ARRL), 225 Main Street, Newington, Connecticut 06111, is an organization that represents ham radio operators throughout the United States. The ARRL can direct you to the ham club in your area. The following magazines are primarily of interest to ham radio operators, but many of the ideas and concepts can also be applied to the hobby of scanning: *CQ Magazine*, 76 N. Broadway, Hicksville, NY 11801, and *73 Amateur Radio Today*, P.O. Box 7693, Riverton, New Jersey, 08077-7693.

Listening and transmitting

Did you know that you can become a licensed ham radio operator? The new no-code license has eliminated the need to learn Morse code. The actual test consists of a few multiple-choice questions that are not difficult to answer. If you want to transmit on the air as well as listen, contact the ARRL for additional information.

Frequency suppliers

At this writing, there are two companies that provide frequencies on computer disks: Grove Enterprises, P.O. Box 98, Brasstown, NC 28902, and PerCon Corporation, 4906 Maple Springs/Ellery Road, Bemus Point, NY 14712. PerCon also provides the complete Federal Communications Commission (FCC) data base on CD ROM. The FCC data includes the frequency assignments for every licensed transmitter in the United States! Police, fire, ambulance, business radio, and hundreds of additional agencies are listed in the data base. The data is so extensive that it would fill a room with volumes of books.

Frequency management software

Storing, sorting, and printing thousands of frequencies is made possible by a variety of computer programs. One of the popular programs is *Radiolog,* a program that is available from shareware distributors. As you probably know, shareware programs are available at budget prices. If you like the program and decide to use it, you send a registration fee to the author. Shareware dealers are listed in the classified section of computer magazines.

Computer controlled scanning

The technology of the nineties has made it possible to link your scanner to a computer. The computer automatically logs and stores each frequency that is received by your scanner radio. The specially designed computer interface and software makes it possible for your computer to baby-sit your scanner for an hour, or for several days! New frequencies are easily discovered and automatically stored in the computer's memory. For additional information, contact the following suppliers: Datametrics, Inc., 2575 South Bayshore Dr., Suite 8A, Coconut Grove, FL 33133; Computer Aided Technologies, P.O. Box 18292, Shreveport, LA 71138; and Commtronics Engineering, P.O Box 262478, San Diego, CA 92196.

Building a library

As you begin to build a scanning library, you'll quickly discover that there is a wide variety of scanning publications. Computers and desk-top publishing software has made it possible for anyone to produce a scanning newsletter or bulletin. Some of the publications are well written and informative, others are not. Make your selections wisely and don't forget to utilize common sense.

11

CHAPTER

Questions and answers

As you become familiar with the hobby of scanning, your curiosity and hands-on experience will undoubtedly produce a wide variety of questions. The most commonly asked questions are listed below. For your convenience, the questions are grouped alphabetically and under specific titles.

Antennas

Q *What is the best scanning antenna?*

A If you're only interested in local monitoring (25 miles or less) nearly any antenna will suffice. To receive radio communications from greater distances, say 25 to 100 miles, a ground plane or directional beam would be a good choice.

Q *Are the more expensive antennas capable of receiving more signals?*

A Generally speaking, the elements of a quality antenna will be thicker, more corrosion resistant, and capable of producing a superior signal.

Q *What is the ideal height of an outside antenna?*

A The antenna should be mounted above the surrounding structures. If that's not practical, simply mount the antenna at a reasonable height (3 to 6 feet) above the supporting structure.

Q *Should outside connections be waterproofed?*

A Absolutely! Radio Shack coax sealant (catalog #278-1645) is highly recommended.

Q *Do outside antennas require periodic maintenance?*

A Yes. The antenna should be disassembled and cleaned every two years. Many hobbyists apply a thin coat of shellac or urethane to the element contact areas. The coating retards corrosion and prolongs maintenance intervals.

Q *I live in an apartment and can't erect an outside antenna. Will an indoor antenna work as well?*

A Indoor antennas don't usually work as well as outside antennas. For the best indoor reception, try the antenna in several locations and check the signal strength.

Q *Can I use a commercial grade 800 megahertz transmitting antenna to monitor the 800-MHz band?*

A Yes. Any good transmitting antenna will make a good receiving antenna.

Q *My antenna uses a* balun. *What does it do?*

A The maximum signal transfer from your antenna to receiver occurs when the antenna impedance matches the receiver input impedance. If the system is mismatched, signal strength will be compromised. The balun is a matching transformer that helps to provide maximum signal transfer between your antenna and feed line.

Q *Will grounding my scanning antenna improve its performance?*

A No. Antennas are grounded to help protect your home and equipment from lightning and static discharge. The procedures for grounding your antenna will probably be regulated by local building codes. Insurance companies can refuse to settle a claim if your grounding procedures did not comply with state and local code requirements.

Coaxial cable

Q *What type of cable should I use for scanning?*

A RG-6 cable, which is commonly available, is the ideal choice. RG-6 provides low loss at a reasonable price.

Q *How often should coax be replaced?*

A Coaxial cable that has been exposed to the weather should be replaced at least every four years. Coax that has been exposed to extreme weather conditions should be inspected yearly for cracks and abrasions. Damaged cable should be replaced immediately.

Q *How can I match a connector with a specific cable?*

A Each brand of coax cable, RG-6, RG-8, RG-58, etc., requires a different connector. Commercially available connectors are packaged by cable size. An RG-8 cable connector won't work on RG-58 cable. Read the labeling before you buy.

Q *Why are there so many different types of coaxial cable?*

A Different frequencies and different applications require special coax. The problem becomes especially important if the cable is used for transmitting purposes. The coax selected must be capable of handling the power output of the transmitter. Generally speaking, large diameter cables can handle more power than small diameter cables. However, cable diameters can't be so large as to restrict handling and ease of concealment. All of these considerations and more require special cables of various sizes.

Listening laws

Q *If I listen to cellular phone calls on my TV set, am I breaking any law?*

A Yes. The Electronic Communications Privacy Act prohibits cellular monitoring on any receiver.

Q *What is the cutoff date for the manufacture of scanner radios that can receive the cellular phone frequencies?*

A Cellular-capable scanner radios cannot be manufactured or imported into the U.S. after April 26, 1994.

Q *Will it be lawful for individuals to own and resell their used cellular-capable scanner radios after April 26, 1994?*

A Yes. The law only applies to new equipment that is manufactured after the cutoff date.

Q *Are wideband, 800-MHz converters affected by the new listening laws?*

A Yes. The law bans any converter that can tune the cellular phone frequencies. Because it is nearly impossible to design an 806- to 960-MHz converter that omits cellular coverage, all 800-MHz converters will be unlawful.

Q *What services are off-limits to monitoring?*

A Scanning laws forbid the monitoring of encrypted or scrambled signals, paid subsidiary carriers (FM radio stations that operate without commercials, usually heard playing in department stores), mobile/cellular phones, voice paging, and remote broadcast links (used by reporters to communicate with news room).

Name that frequency

Q *What are the frequencies used by the new 900-MHz cordless phones?*

A Search between 902.00 and 928.00 MHz. The exact frequencies vary between model and manufacturer. When you locate an active frequency, add it to your scanner radio's permanent memory.

Q *Where can I find hotel frequencies?*

A Hotel frequencies are widely scattered in the 151.00-, 154.00-, 463.00-, and 464.00-MHz ranges.

Q *What is the primary nationwide repeater frequency for the Secret Service?*

A The nationwide repeater frequency is 165.375 MHz.

Q *What are the frequencies used by the old, conventional mobile phones?*

A Prior to cellular phones, it was possible to make phone calls through two-way radio equipment. The frequencies are still in use and can be located by searching through the following frequencies: 152.030 to 152.210, 152.510 to 152.810, and 454.025 to 464.650 MHz.

Q *What are the frequencies used by radio-controlled toys?*

A If you enjoy listening to odd sounds (no voice will be heard), here are a few of the frequencies that are utilized: 26.995, 27.045, 27.195, and 27.255 MHz.

Reception problems

Q *I am hearing police transmissions on 133.985. As you know, this is the aircraft band. How is this possible?*

A You are hearing an image from the 155.00 megahertz police band. The actual police frequency will be 21.40 or 21.70 MHz above the aircraft band frequency.

Q *How can I convert a scanner radio to receive the 800-megahertz band?*

A If your receiver doesn't have at least part of the 800-MHz band, it can't be modified to receive specific 800-MHz frequencies.

Q *I can only hear one side of cordless phone conversations. Why?*

A You're listening to the handset frequencies in the 49.0-MHz range. To hear both sides of a cordless conversation, monitor the 46.0-MHz frequencies.

Q *When I enter 162.4375 into my scanner, it appears as 162.435. What happened?*

A Scanner radios typically have 5-kHz resolution. That's not very accurate, but since narrowband FM signals are very broad, you'll hear the station.

Q *Will the aluminum siding on my house affect my indoor scanning antenna?*

A All metallic surfaces will reflect radio waves. If you can place your indoor antenna near a window, reception might be satisfactory.

Scanning equipment

Q *What is the best scanner radio to buy?*

A It's similar to asking what car should I buy. Your selection should be based primarily on frequency coverage, the number of available memories, and cost. If you don't need expanded frequency coverage and can live with 20 channels or less, you can select from several models that are priced between $80.00 and $150.00. Scanner buffs interested in 800-MHz coverage, military air frequencies, and high memory capacity, will need to select a scanner radio that is priced between $250.00 and $400.00.

Q *What are the names of the top scanner radio manufacturers?*

A Programmable scanner radios are available from AOR, ICOM, Kenwood, Radio Shack, Uniden-Bearcat, and Yaesu.

Q *What is "dynamic range?"*

A The ability of a scanner radio to receive both weak and strong signals, without causing overload problems. Since all scanner radios have poor dynamic range, it won't be listed in the manufacturer's specifications.

Q *Do cable TV adapters and splitters affect the strength of radio signals?*

A Yes. When you add a splitter or adapter, you'll decrease the signal between 30 and 50 percent. In the 800-MHz band, the loss might even be greater. To receive the maximum signal, use a continuous length of coax between your antenna and scanner radio.

Q *Will a preamplifier increase my listening range?*

A If your lead-in is 50 feet or less and if your main listening targets are the public service bands, a pre-amp will do little to improve or increase your reception. Broadband preamplifiers amplify all the signals, including those that you don't want. The result is an increase in images and intermod. If you must use an amplifier, choose one with a narrow bandwidth and low noise characteristics.

Q *Will I void the warranty if I modify my scanner to receive the cellular bands?*

A Probably so. Any manufacturer would certainly frown upon aftermarket modifications. The situation becomes especially tricky if you purchased an extended warranty. If the company detects tampering, they could legally refuse to honor the warranty.

Q *Are there any speech decoders on the market?*

A The Electronics Communication Privacy Act has made decoders illegal to manufacture or use. Even simple speech inversion descramblers are not available.

Q *How can I boost the audio of my handheld scanner when I use it in my vehicle?*

A Plug your scanner radio into Radio Shack's *Compact Disc Cassette Adapter,* catalog #12-1951. The CD adapter plugs into your scanner radio and is inserted into your cassette player. With this setup, pedestrians walking on the street will be able to hear your scanner radio!

Spectrum rules and regulations

Q *I monitored a taxi cab using 470.40 megahertz. How can I find the operating frequency of the base unit?*

A Between 450.00 and 470.00 MHz paired frequencies are separated by exactly 5 MHz. The base can always be found on the lower frequency. By subtracting 5 MHz from 470.40 MHz, the base operating frequency will be found on 465.40 MHz.

Q *What are the basic band plans?*

A The radio bands are separated as follows: 30 to 50 MHz, VHF Low Band; 72-76 MHz, VHF Mid Band; 108 to 174 MHz, VHF High Band; 225 to 512 MHz, UHF Band; and 806 to 960 MHz, microwave mobile or cellular band.

Q *I recently heard about the new Personal Communications Service that will offer long-range cordless phones. What are the frequencies that will be utilized?*

A The new Personal Communications Service (PSC) has been assigned to frequencies between 1850.00 and 2100.00 MHz (1.85 to 2.1 GHz).

Q *Is the FCC planning to reduce the radio spectrum's band spacing to make more room for additional frequencies?*

A Yes. The process, which will take several years, will not begin before 1996. Your scanner radio will still be capable of receiving the new frequencies.

Here is your hobby—scanning!

Your journey through the invisible world of scanning is complete. Your future adventures, however, are just beginning. As you have learned, the hobby of scanning contains thousands of adventures to explore. By simply pressing a button you can experience real-life thrills as they happen. No adventure on land, sea, or air can escape you. You hear everything, you know everything, but you never leave the safety of your favorite chair. It's magical, it's futuristic, and it's your new hobby. Enjoy the adventure!

Glossary

accessory cable Short length of cable with connectors.

attenuator A device that reduces signal strength.

balun Device used for matching a transmission line to an antenna.

band allocations Group of specific frequencies.

beam antenna High gain, directional antenna.

base station Fixed transmitter.

base station scanner Scanner radio designed for tabletop use.

birdies Internally produced harmonics.

CB Citizen's Band radio.

coax Special cable designed to carry radio signals.

continuous coverage receiver Receiver that can monitor specific segments of the radio spectrum without frequency gaps.

digital transmission Computer encrypted radio signal that cannot be monitored.

drip loop Cable configuration that prevents moisture from entering into a structure.

dB switch Limits or increases scanner sensitivity to an incoming signal.

duplex Radio signal that permits both parties to talk simultaneously. (Cellular phones use a Duplex signal.)

ECPA (Electronic Communications Privacy Act) Law that prohibits cellular monitoring.

FCC (Federal Communications Commission) Agency responsible for regulating the radio spectrum.

feed line Cable that connects equipment to outside antenna.

filter A device that blocks a specific frequency or group of frequencies.

frequency spacing Separation between frequencies.

ground plane antenna Omnidirectional, fixed antenna.

ground waves Radio waves that travel near the ground.

ground wire Connects antenna or equipment to ground (earth).

horizontal polarization Antenna elements are horizontal to the earth.

images Duplicates of the original frequency.

impedance Electrical balance between antenna and coax cable.

intermod When strong signals mix together producing a garbled signal.

indoor antenna Any antenna located inside a building.

ionized gas Atmospheric gas that reflects radio waves.

LCD (Liquid Crystal Display) The information window on a scanner radio.

lightning arrestor Device that prevents static build-up in antenna feedline.

line of sight The height of an antenna relative to natural and man-made objects.

longwire antenna Random length of wire used as an antenna.

microvolt One millionth of a volt.

microwave mobile band Commonly referred to as 800-megahertz band.

mobile repeater Vehicle repeater used to amplify low power signals from hand-held transmitters.

mobile scanner Scanner radio designed to be installed in a vehicle.

noise Unwanted disturbance that accompanies the desired signal.

outdoor antenna Any antenna that is exposed to the weather.

portable scanner Handheld, battery operated receiver.

pre-amp Signal amplifier.

repeater Transmitter that receives, amplifies, and re-transmits a radio signal.

rotor Electric motor used to rotate beam antennas.

sensitive communications Radio communications vital to local, state, or national security.

simplex Radio signal that permits user to talk or listen, but not both.

skip Radio signal that travels a considerable distance.

splitter Cable accessory that provides multiple access from a single cable.

tower Antenna supporting structure.

UHF band (Ultra High Frequency) Frequencies between 406.00 and 470.00 MHz.

UHF-T band (Ultra High Frequency) Frequencies between 470.00 and 512.00 MHz.

VHF High Band (Very High Frequency) Frequencies between 138.00 and 174.00 MHz.

VHF Low Band (Very High Frequency) Frequencies between 29.70 and 50.00 MHz.

vertical polarization Antenna elements point toward the sky and ground.

voice security Radio communications that are encrypted.

Index